Ecological effects of waste water

Ecological effects of waste water

E. B. WELCH

Professor of Applied Biology
Department of Civil Engineering
University of Washington
Seattle, Washington

with Hydrographic Characteristics by

T. LINDELL
National Swedish Environmental Protection Board
Limnological Survey
Uppsala, Sweden

CAMBRIDGE UNIVERSITY PRESS

CAMBRIDGE

LONDON   NEW YORK   NEW ROCHELLE

MELBOURNE   SYDNEY

Published by the Press Syndicate of the University of Cambridge
The Pitt Building, Trumpington Street, Camridge CB2 1RP
32 East 57th Street, New York, NY 10022, USA
296 Beaconsfield Parade, Middle Park, Melbourne 3206, Australia

First published 1980

Printed in the United States of America
Typeset by Photo Graphics, Inc., Baltimore, Maryland
Printed and bound by the Murray Printing Company,
Westford, Massachusetts

Library of Congress Cataloging in Publication Data

Welch, Eugene B.

Ecological effects of waste water.

Includes bibliographical references.

1. Sewage – Environmental aspects.   I.   Lindell, T.
II.   Title.
QH545.S49W44   574.5'263   78-11371
ISBN 0 521 22495 0 hard cover
ISBN 0 521 29525 4 paperback

To my wife, Karen, whose positive, steady, and generous support is a critical ingredient in all my endeavors, including this one – and to the late Helen Burton, my dear aunt, who was so fond of the printed page.

CONTENTS

vii

PREFACE

The study of applied ecological concepts in relation to the impact of man's wastes on aquatic environments should be an integral part of any environmental engineering curriculum. A general understanding of such concepts is necessary not only for biologists, but also for engineers, chemists, and other nonbiologists, because planning and implementation for water-quality control requires knowledge of the complete process, from problem identification and an understanding of the cause(s) to an analysis of alternatives. For example, one should recognize the quality level of an environment, the degree to which that quality departs from a natural level, the relative effect of any waste(s), and the degree and rate of improvement if wastes are reduced.

Water-pollution control has gone through an interesting evolution – from one of benign neglect to the institution of minimum controls when the cause and level of severity could be demonstrated by the plaintiff and finally to a hard-line emphasis on polluters rather than the water. With the water-pollution act of 1972 and common-law cases such as the State of Illinois versus the City of Milwaukee for contamination and eutrophication of Lake Michigan, it may seem to some that an accurate analysis of ecological cause and effect is becoming unnecessary. Actually, however, it is becoming more and more necessary and is compounded by the need to be able to predict the impact resulting from changes in waste input. The original problem was to demonstrate the amount of damage; now the question is more sophisticated: How much improvement will occur and how soon?

The course upon which this text is based, which is offered by environmental engineering conjoined with fisheries at the University of

Washington, has been very successful in drawing seniors and graduate students from biological as well as nonbiological curricula. By emphasizing the synecological approach to practical pollution problems, via a rather thorough analysis of several water-quality variables, the text challenges biologists and nonbiologists alike. Other texts on water-pollution biology have inadequately discussed plankton and periphyton and the causes and control of eutrophication. Plankton is emphasized in this text, but the other more normal components, such as benthos and fish, are also included. The value of this text is breadth, appealing to biological and nonbiological students alike, with its greatest appeal directed to interdisciplinary programs.

Aquatic ecology is a rapidly advancing field and many of the concepts covered in this text are as yet not completely understood. The student and professional reader should not consider any writings in this field, including this one, as the last word, but only as a point of departure for further learning. In this view it is my hope that the student's appetite is whetted for more knowledge of aquatic ecology and the effects of wastes, which in the long run should lead to better environmental protection.

I am indebted to many who have assisted in countless ways in the development of this text; to my colleagues and students, too numerous to mention individually, who have contributed data and ideas, to my friends and colleagues R. C. Averett and I. and G. Ahlgren who reviewed the manuscript, and especially to T. Lindell, who contributed the chapter on hydrographics, and to R. E. Nece, who reviewed it.

E.B.W.

INTRODUCTION

The subject of waste-water effects on ecological systems encompasses great breadth and variability. Dissecting ecological systems into easily studied physical units or individual populations and isolating the effects of myriad wastes, one at a time, results in few generalizations about all wastes in all aquatic ecosystems. A better method is to examine each type of waste individually in relation to representative groups of organisms in a shopping-list approach. A still better way is to study the functioning of ecological groupings of aquatic populations within the abiotic environment and then to study each group's functioning with respect to a limited variety of wastes. The last approach is used in this book because it emphasizes consideration, first, of the natural function of the particular biotic component, and then of the effect of the waste(s) most pertinent to that group. The ecological groups discussed include the phytoplankton, zooplankton, periphyton, macrophytic rooted plants, benthic macroinvertebrates, and fish (the last mostly as a bioassay animal and as a target for standards of water quality).

Obviously, if all these groups are to be considered, the limitation must be on the variety of wastes discussed. The limitation is handled to some extent by considering the effects of wastes most important to each group and those waste constituents most closely associated with the natural processes and their control. For example, nutrient elements are extensively discussed with respect to the algae, both planktonic and periphytic, because these groups react first to an increased supply of limiting nutrient. On the other hand, a lowered concentration of dissolved oxygen does not greatly affect the algae because of their autotrophic nature and is not discussed in relation to them. However, the dissolved oxygen level is an important factor in regulating the response of macroinvertebrates and fish to oxygen-demanding wastes and is, therefore, considered in relation to these animals.

Selecting the most pertinent (in my opinion) wastes with respect to each functional group has necessitated slighting certain wastes. Nutrients, heat, and organic waste are discussed in considerable detail; toxicants are given less attention. The toxicity limits of specific compounds are considered in some detail with respect to fish, and the relative effect of "toxicity" in general is stressed in connection with benthic community composition.

This approach dwells on the concepts relating to the effects of particular wastes on community structure and processes, with only minor emphasis on any given species. Admittedly, information is lost when individual species populations are not considered, and not all species in a genus or family react similarly to a waste. However, available information is not adequate to allow analysis of the effects of wastes on individual species (except in the case of some fish). Further, information on community structure and processes is usually adequate to permit description of the ecological effects of a waste and is more meaningful to persons charged with the management of water resources than are lists of species and their quantitative changes. The most useful goal from the standpoint of management is to describe changes through measurements of structure (e.g., species diversity and biomass, noting any changes in dominant species) and of processes (e.g., productivity, respiration rate, growth rate, and reproductive rate). Of course, not all measurements are equally appropriate to all groups of organisms, and that is a principal point that is discussed.

A significant goal of this book is to convey an understanding of how the functional groups of organisms respond, first to natural factors and then to the superimposition of representative wastes. But more importantly, given that information, the next step is to sort out the important controlling factors and to predict the direction and in some cases the relative magnitude of change.

Chapters 1 through 6 present some general concepts of ecology and explain how they are related to the management of ecosystems. These concepts represent a working philosophy about aquatic systems that is a necessary background for the discussion in Chapters 7 through 12 of specific wastes and their effects.

PART ONE

GENERAL CONCEPTS OF AQUATIC ECOLOGY

1

AQUATIC ECOSYSTEMS AND MANAGEMENT

An *ecosystem* can be described as some unit of the biosphere, or the entire biosphere itself, within which chemical substances are cycled and recycled while the energy transported as part of those substances continually passes through the system. Although every ecosystem must ultimately obey the laws of thermodynamics and degrade to complete randomness (*wind down*), consistent with the universe, energy may be accumulated momentarily in an ecosystem. However, without a continuously renewed input of energy, the accumulation would be exhausted and the system would wind down.

The single continuous input to the world's ecosystems is from solar energy, and the conversion of that electromagnetic energy into chemical energy and then into work is what allows an ecosystem and the organisms that make up that system to function.

Because the processes of energy flow and nutrient cycling are quite variable, the assignment of clear boundaries between what might otherwise seem like clearly separate systems is not easy. However, for the sake of practicality and manageability, boundaries are cast that allow ease of study and process measurement. Thus, a stream and its immediate watershed as well as a lake and its watershed inputs are considered ecosystems. The system could be considered closed under some conditions if one is describing chemical nutrients, but never if one is referring to energy, because there is always an input to and a loss from the system. Because an ecosystem responds to inputs as an integrated system, the study of whole systems is useful for management.

Ecosystem composition and energy sources

Each ecosystem has a structure that determines how it functions in the transfer of energy and the cycling of nutrients. This structure can be thought of as the organization of the internal groupings of chemical nutrients and energy through which the functioning occurs. This matter is living and dead, the living being represented best by the trophic levels of organisms leading from algae and/or rooted plants to fourth-level carnivores, for instance, in some grazing food webs. Outside organic matter is processed and decomposed by insects, fungi, and bacteria through the carnivores in detritus-based food webs.

The watershed has come to be regarded as an integral and insepar-able part of the ecosystem (Borman and Likens 1967; Likens and Bor-man 1974). The character of the watershed determines whether the stream's energy source is mostly *autochthonous* (produced within) or *allochthonous* (produced outside). In a comparative study of the North-east area, Likens (personal communication) has shown that an auto-chthonous-producing forest results in the domination of the drainage stream by allochthonous inputs, whereas the watershed lake is auto-chthonous. Wissmar found a high mountain lake in a coniferous forest dominated by allochthonous inputs (Table 1). Lakes and streams in poorly vegetated watersheds and relatively large lakes tend to be dom-inated by autochthonous sources, as do also lower stretches of streams and large rivers (if not turbid).

It is necessary to know the proportions of allochthonous to authoch-thonous energy sources in ecosystems because of the management decisions affecting aquatic ecosystems that must be made. To effec-tively protect and use aquatic systems, man must know how systems use and respond to different amounts and varying compositions of natural energy.

Table 1. *Comparison of energy sources in two ecosystems, in grams of carbon per square meter per year*

Energy source	Deciduous forest			Coniferous forest – lake
	Forest	Streams	Lake	
Autochthonous	941	1	88	4.8
Allochthonous	3	615	18	9.4

Source: Data on deciduous forest from G. E. Likens, personal communication; data on coniferous forest from Wissmar et al. (1977).

Energy flow and nutrient cycling

If one could measure all the organisms and their main processes in an ecosystem and if their energy-consuming and energy-processing characteristics fell neatly into separate levels, one would be able to arrange a flow diagram as shown in Figure 1. Such a diagram shows several important points: (1) energy flows through the system and does not return, because by the second law of thermodynamics matter moves toward randomness, from states of high concentration to states of low concentration, and when that happens the energy contained becomes less (note entropic heat loss); (2) loss in energy as heat occurs at each step in the transfer process; and (3) allochthonous energy moves through the heterotrophic microorganisms or decomposers. A significant role of net production of usable energy for consumers is attributed to these microorganisms (note arrow from decomposers to consumers) in addition to decomposition (Pomeroy 1974).

Even for allochthonous sources from surrounding forests or from man's input of organic waste, the ultimate source must be either photosynthesis or to a lesser extent chemosynthesis.

The process of *photosynthesis* includes two phases: (1) a light reaction that traps solar energy and releases molecular O_2 and (2) a dark reaction (light not needed) that utilizes the trapped energy as ATP (adenosine triphosphate) and fixes CO_2 or HCO_3 into cell material. The process yields energy and synthesizes new cells and can be summarized by:

Figure 1. Energy flow and nutrient cycling in an aquatic ecosystem (boundary indicated by dashed line). CB, chemosynthetic bacteria; PC, primary consumers; SC, secondary consumers; TC, tertiary consumers.

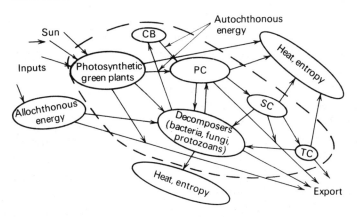

$$6\ CO_2 + 12\ H_2O + \text{light} + \text{chl } a \text{ and accessory pigments} \rightarrow$$
$$C_6H_{12}O_6 + 6\ O_2 + 6\ H_2O$$

The organic compound is glucose – the compound that holds the sun's energy trapped by photosynthetic organisms and allows the maintenance of our biosphere (that thin film of atmosphere, water, and land around the planet Earth that supports life). The process is performed by green plants and a few pigmented bacteria. Only the green plants release O_2 as a by-product.

Chemosynthesis is the other process through which organisms are totally self-sufficient in trapping energy and building cell material. This process yields energy through the oxidation of reduced inorganic compounds and thus requires no previous biological mediation by the various bacteria utilizing this process. The primitive earth was rich in reduced inorganic compounds that are currently used by some bacteria for energy. An example is nitrification:

$$2\ NH_3 + 3\ O_2 \rightarrow 2\ HNO_2 + 2\ H_2O + \text{energy}$$

This process is less important to ecosystems as an energy yielder and more important as a material recycler, because chemosynthetic processes by bacteria are heavily involved in the cycling of such nutrients as nitrogen and sulfur.

Nutrients, such as N, C, P, and S, follow the same pathways as energy in the system; they are consumed by autotrophs (green plants and chemosynthetic bacteria) from inorganic pools and fixed into organic compounds. These nutrients are then transferred through trophic levels just as energy is, because the reduced organic compounds are the carrier of the entrapped chemical energy. For example, CO_2 is the most oxidized state of C, but when fixed into glucose through photosynthesis, 1 mole ($C_6H_{12}O_6$) contains 674 kilocalories (kcal) of energy releasable through respiration by plants, animals, or decomposer microorganisms by the same biochemical pathways. The energy content of whole organisms in the various trophic levels ranges from 4 to 6 kcal gm^{-1} dry weight.

The principal difference between energy and nutrient transport is that once organisms have utilized the energy from complex compounds and oxidized them completely (e.g., glucose to CO_2 and H_2O), the compounds are recycled through the inorganic pool(s) and are almost totally reusable by the community. There is no permanent loss, but, practically, a certain fraction may be lost to the sediments and require tectonic uplift for recycling. The recycling process is shown in Figure 2 (note heavy arrow from decomposers to inorganic pool).

Efficiency of energy and nutrient use

The efficiency of the transfer of energy and/or nutrients is usually measured by the ratio of net productivity of a trophic level to the net productivity available for its consumption (Russell-Hunter 1970, pp. 25, 191). These values usually range between 10% and 20%. The level of this efficiency depends greatly upon the structure of the food web through which the materials are moving.

Structure in ecosystems can be thought of as the organization of species populations into appropriate trophic levels. However, trophic level is rather artificial and few organisms conform to a single trophic level throughout their entire life cycle. Nevertheless, populations organize, in time, in such a way that energy usage is optimized. Ecosystems that are physically stable, such as tropical rain forests and coral reefs, maximize organization and complexity and remain rather constant in biomass, productivity, species diversity, and, consequently, efficiency. Very simply, the greater the variety of energy users in an ecosystem (alternate pathways), the greater the chance that a quantity of energy packaged in a particular way will be intercepted and used before it leaves the system. Instability in ecosystem structure brought about by natural or man-caused variability in the physical-chemical environment results in decreased efficiency and instability in energy and nutrient usage, and nutrient recycling is, therefore, the "looser." Where systems approach steady state, that is, inputs equaling outputs, nutrient recycling tends to be "tighter" and "more complete" (Borman and Likens 1967; Bahr et al. 1972).

Figure 2. Transport and recycling of nutrients through an ecosystem (boundaries indicated by dashed line). CB, chemosynthetic bacteria; PC, primary consumers; SC, secondary consumers; TC, tertiary consumers.

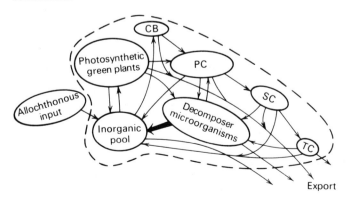

Odum (1969) has illustrated the effect of maturity on the characteristics and processes of ecosystems (Table 2 and Figure 3). He contends
that diversity tends to be low in immature systems and high in mature
systems. Correlated with diversity is relative stability, although how
to define and measure stability is still a matter of controversy. However, net production tends to be inversely related to diversity, and
Margalef (1969) has suggested that at low diversity, fluctuations in
production per unit biomass are quite large and diversity changes are
small, suggesting instability, whereas at high diversity, fluctuations in
production per unit biomass tend to be small with considerable
changes in diversity, indicating stability. One can see from Figure 3
that with time ecosystems tend to develop into a rather steady-state
condition, with little net productivity and maximized structure. If production is low and associated fluctuations are small in a mature ecosystem, it follows that nutrients are conserved and cycled more slowly
and energy thus moves through the system at a slower rate and entropy
is less – the system is more efficient in its functioning.

Examples of efficiency in ecosystems give some idea of their maturity. Fisher and Likens (1972) showed that Bear Brook, a northeastern
forest stream, was only 34% efficient (total respiratory loss/total energy
input). Thus, 66% of the energy left the system downstream unused,
indicative of immaturity. Lindeman's (1942) study of Cedar Bog Lake

Figure 3. Conceptual plot demonstrating successional changes in
species diversity, gross production, and biomass. The shaded area
representing net production approaches zero as the ecosystem
matures. (Bahr et al. 1972, modified from Odum 1969)

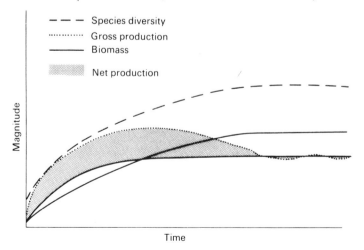

showed it to be 54.5% efficient with 45.5% lost to sediments. Odum's (1956) results from Silver Springs in Florida showed only 7.5% lost from the system downstream with 92.5% being utilized.

Management of ecosystems

What does man want from aquatic ecosystems? Obviously he does not want to destroy them, but he is highly inclined to manipulate them to his purpose whether it be to increase or decrease their productivity, keep them natural, or have them accept (assimilate) wastes with a minimum adverse effect on other uses. The problem is that these uses are conflicting. For example, to produce more and bigger sport fish could require stimulation of primary production, which, in turn, will result in decreased structure (species diversity) and less stability. The yield of particular game-fish populations may increase, but at the expense of the overall stability and efficiency of energy-handling and nutrient-recycling capabilities of the system. In the same light, ecosystems cannot assimilate wastes without some cost to their structure and stability, with consequent reduced efficiencies in nutrient cycling and energy utilization. Instability may produce a more variable oxygen content, which could reach lethal limits at times.

Stable communities are probably not more able to resist change from waste input than unstable ones. Bahr et al. (1972) argue that if this were true, species diversity would decrease with waste input (abiotic change) at an increasing rate (curve B in Figure 4), rather than at a decreasing rate as is usually observed (curve A). Highly diverse com-

Table 2. *Characteristic properties of ecosystems*

Property	Type of ecosystem	
	Imma-ture	Mature
Net production (yield)	High	Low
Food chains	Linear	Weblike
Nutrient exchange	Rapid	Slow
Nutrient conservation	Poor	Good
Species diversity	Low	High
Stability-resistance to perturbations	Low	High
Entropy	High	Low

Source: (Odum 1969), with permission of the American Association for the Advancement of Science.

Figure 4. Ecosystem stability as a function of intensity in a stressful abiotic change. The two curves are discussed in the text. (Bahr et al. 1972)

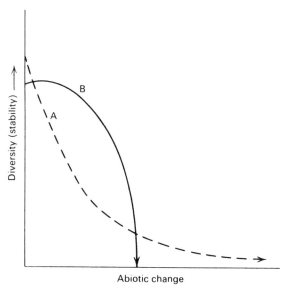

munities should produce stable behavior in ecosystems, but such a state is very sensitive in itself to change in the environment. Any change, or manipulation, tends to lower that stability and decrease the efficiency of ecosystem functioning. Man probably cannot have his cake and eat it too; he cannot maintain highly structured and diverse ecosystems resistant to change and at the same time maximize their recycling and energy-using efficiency.

2

STANDING CROP AND PRODUCTIVITY

Before population growth and control are discussed further, some important terms should be understood. *Standing crop* is often synonymous with biomass in that it represents the quantity per volume or area that is present and can be measured. The units of measure are wet weight, dry weight, numbers, or the quantity of some elemental constituent (N, P, C). Of course, standing crop is the more general term; biomass is appropriate only when the unit is mass.

Productivity is the rate of biomass formation and is a dynamic process. Productivity is often proportional to biomass, but in the case of fast-growing and efficiently grazed populations, biomass may be very small, whereas daily productivity is very high. To understand that phenomenon, the concept of *turnover rate* is useful and can be defined as:

$$\frac{\text{productivity } (ML^{-3}T^{-1})}{\text{biomass } (ML^{-3})}$$

where M = mass, L = length, and T = time.

Turnover rate is thus the number of times that the biomass is theoretically replaced per unit time. Because they have the same units (T^{-1}), turnover rate equals growth rate but only when biomass remains constant. When biomass is not constant from one time to another, the growth rate over that period is determined by the increase or decrease in biomass as well as the instantaneous measurement of productivity.

Pyramid of biomass

Productivity is directly proportional to biomass in a food chain only if the turnover rate at each trophic level is constant. An illustration of this is shown in Table 3. Because energy is lost at each transfer in the food chain, productivity must always decrease from primary producer to secondary consumer. In this instance, productivity decreases by a factor of 10 at each level. There is no reason to believe turnover rate would remain constant. Biomass may be greater at consumer than producer levels because producers are smaller organisms with a higher ratio of cell surface to volume and therefore result in more rapid turnover rates. In such a case the biomass pyramid may be inverted as cited by Odum (1959, p. 64). In Long Island Sound zooplankton and bottom fauna biomass were double that of phytoplankton. In the English Channel the ratio was 3:1. Thus, from this theoretical standpoint, productivity may not, in fact, be very tightly related to biomass in nature, particularly in open water systems. If the biomass in the examples from Odum was multiplied by turnover rate, the pyramids of biomass would be pyramids of productivity (energy flow) and become upright.

Pyramid of nomenclature

Primary production is the rate of biomass formation of the autotrophs, which are primarily the green plants. *Net production rate* is measured as the net amount of organic matter fixed and transferrable to the next trophic level; *gross production* equals total assimilation including respiration. These quantities can be represented as shown in Figure 5.

Efficiency of transfer

The energy efficiency of ecosystems has direct management implications, as pointed out earlier. At the first trophic transfer, green plant photosynthesis is rather inefficient. For terrestrial systems, efficiency of light utilization is about 1% and in water it is about 0.1–

Table 3. *Hypothetical relation of biomass and productivity in a three-tiered food chain. [Units are in mass (M), length (L), and time (T)]*

Trophic level	Biomass (ML^{-2})	Productivity $(ML^{-2}T^{-1})$	Turnover rate (T^{-1})
Secondary consumer	1	10	10
Primary consumer	10	100	10
Primary producer	100	1000	10

0.4%. The lower efficiency in water results from the absorptive and scattering effects of water and its particles. Utilization of primary productivity by secondary consumers is about 10% (range from 5-20%). The 90% that is lost at each trophic transfer is in the form of heat. In essence, the efficiency of energy utilization at each trophic level depends upon how individual consumers handle the energy ingested (Figure 6). Trophic transfer is relatively efficient or inefficient because of the individual populations that occur there. If the structure of an ecosystem is disrupted by waste input such that many of its efficient energy utilizers are eliminated, then overall efficiency in the system would naturally decrease. Thus, degradation of community structure also theoretically has a degrading effect on the dynamics of the system.

Figure 5. Hypothetical diagram of a simplified ecosystem energy budget. P, gross production; NP, net production; R_{SA}, respiration autotrophs; R_{SH}, respiration heterotrophs; NEP, net ecosystem production. (Modified from Woodwell 1970)

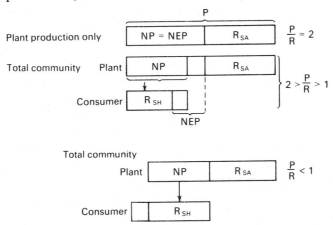

Figure 6. Energy budget of individual animal. PB, plant biomass; TI, total energy ingested; TA, total energy assimilated; NA, nonassimilation (excretion); RA, respired assimilation (maintenance and activity demand); NRA, nonrespired assimilation; NTG, nontrophic growth; NP, net production. (After Russell-Hunter 1970)

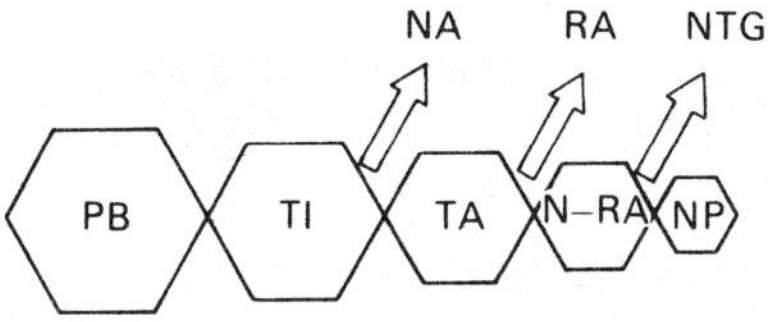

3

HYDROGRAPHIC CHARACTERISTICS

The effects of waste water on a certain environment will, to a large extent, be determined by the physical characteristics of the receiving water. Sometimes it is possible to modify the initial condition for the mixing process by, e.g., adding the waste water at a certain convenient level or using different types of diffusors. The final concentration of a contaminant in the receiving water is thus determined by these initial manipulations plus the physical characteristics of the waste water and receiving body. One type of environment, effective in dispersing the contaminant, may produce low concentrations; another environment, capable of very little dispersion, may produce high concentrations.

Also, the processes of sedimentation are primarily a function of the physical characteristics of the water. The study of the movement of water particles is, therefore, fundamental to the understanding of the processes present in the water environment.

Types of flow

Depending upon the movements of water particles, the flow may be either *laminar* or *turbulent*. Within a laminar flow the water movement is regular and horizontal, that is, the separate layers flow on top of each other without extensive mixing. The mixing that does occur is on the molecular level. Laminar flow, however, is a rare phenomenon in nature.

Water movement that is irregular – causing mixing eddies to develop on a macroscopic scale – is called turbulent. Within a turbulent water mass the water velocity varies considerably in both vertical and horizontal directions. The movement is characterized by erratic motions of fluid particles and considerable lateral interchange of momentum. This type of movement is, therefore, most often simply described by the aid of statistical concepts. The turbulent motion is totally dominating in a natural environment.

When flow is constant in both time and space, it is termed *uniform* and is characterized by parallel stream lines. Even though a uniform flow is also very rare in nature, the term is often useful in modeling the processes present. The opposite to a uniform flow is called *non-uniform*.

If the flow varies with time, it is termed *unsteady*; a flow that is at every point constant with respect to time is called *steady*. The steady-state situation is also a very useful generalization in modeling processes. A laminar flow occurs where the water velocity is low and/or close to an underlying solid, smooth surface. In this case the flow is sometimes both steady and uniform. Steady and uniform conditions are rare in nature, but a situation close to this state is often found in near-bottom layers in a lake. The laminar, uniform flow is easily transformed into a turbulent flow with only a small change in velocity. When one layer of fluid slides over another, the friction between the layers gives rise to a shearing stress that with increasing velocity results in a turbulent flow. This state is reached faster if the underlying bottom is rough. The shearing stress for laminar flow is expressed as:

$$\tau = \mu \cdot \frac{du}{dz}$$

where τ = tangential force per unit area, μ = coefficient of viscosity, and du/dz = velocity gradient perpendicular to the direction of motion.

For turbulent motion the eddy viscosity is much larger than the molecular one, and the shear stress is often written as:

$$\bar{\tau} = \epsilon \cdot \frac{\overline{du}}{dz}$$

where ϵ = eddy viscosity.

It is beyond the scope of this chapter to discuss turbulence in detail. However, the effect of vertical turbulence on phytoplankton production should be indicated. Without turbulence to maintain plankton algae in the lighted zone they would sink and productivity would be minimal. This is particularly true for diatoms that have specific gravities of 1.02–1.03. Figure 7 illustrates how turbulence maintains diatoms suspended in the water column. Because cells tend to accumulate at depth the concentration there tends to increase and the concentration near the surface tends to decrease. When this occurs, turbulence tends to transport more cells toward the surface than to depth. Therefore, the effects of sinking tend to be partially offset by turbulence, which redistributes the cells through the water column and maintains cells in the lighted zone. The effect of turbulence as it affects productivity and plankton species succession will be discussed again in Chapter 7.

Regimes of flow

If the water is moving fast enough, as in some rivers, it is often impossible for a (gravity) wave to travel upstream. The velocity of propagation of a gravity wave is

$$c = \sqrt{gh}$$

where g = gravity and h = depth of water.

The flow in this case is termed *shooting*. If the flow is slower than the propagation of a gravity wave, the term *tranquil* (or streaming) is used. It is possible to separate the different regimes of flow according to the water velocity and the distance from a solid river bottom (Figure 8). There is a transition zone between a turbulent and a laminar flow within which the type of flow is essentially dependent on the water temperature.

Boundary layers

Figure 8 shows schematically the flow conditions in a stream within the first few cm above the bottom. Very close to the bottom

Figure 7. Effect of turbulence on the vertical distribution of phytoplankton.

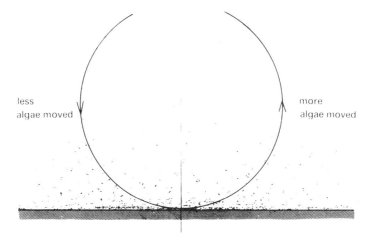

less
algae moved

more
algae moved

there exists a thin layer where the conditions undergo rapid changes. The water in immediate contact with the bottom adheres to it, and there is, therefore, a sharp velocity gradient in this zone. This layer, in which the flow is appreciably retarded by the viscosity of the water and the friction against the bottom and within which there are shear stresses in the fluid, is called the *boundary layer*. The flow in the boundary layer is either laminar or turbulent. In rivers the boundary layer is turbulent and often extends throughout the entire water column as the friction against the bottom is very predominant, and velocity gradients exist throughout the entire depth. This boundary layer could be laminar in channels with a very slow flow or under special conditions in lakes.

The initiation of turbulence in the boundary layer is primarily a function of the water velocity and the stability of the flow (see the following discussion of density effects).

In natural water courses the boundary layer may be upset by the character of the bottom; for example, if the bottom is convex, the flow on the lee side of the convexity tends to create a reverse velocity component close to the bottom and causes a *boundary layer separation* (Figure 9).

The boundary layer separation, the other effects of irregularities of the bottom, and the shear stresses between adjacent water layers for

Figure 8. Regimens of water flow in a wide, open channel. See text for explanation. (Sundborg 1956)

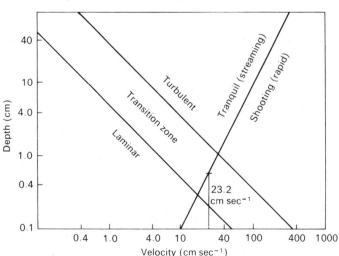

common velocity patterns justify the consideration that almost all movements in natural waters are turbulent. The interaction between the bottom and the overlying water is very important from the standpoint of water quality. The processes of erosion, deposition, and resuspension of sediment are dependent upon the variations in the state of turbulence, the viscous shear, and the form resistance of the bottom.

Density effects

The velocity patterns and the state of flow in regard to the processes of deposition and resuspension are equally important in regard to the dispersion and mixing of a contaminant entering a receiving water. For example, within a slow, near-laminar running stream a contaminant can accompany the stream several kilometers without mixing on a macro scale; however, in a fast-moving turbulent stream the mixing may be practically complete in as little as 10 meters or so downstream, depending on stream size, manner of contaminant introduction, and so on. We have so far in our discussion assumed a water volume of equal density. The conditions mentioned above could, however, be strongly upset by density differences within the water mass. Density differences are very common features of both lakes and estuaries and sometimes of rivers, and therefore the causes and effects of water density must be understood.

Density gradients in a water mass are caused by differences in water temperature and in dissolved or suspended material in the water. In lakes, temperature is the completely dominating cause of density stratification; in estuaries dissolved substances (salinity) often constitute the stratification. In the river environment, suspended material is an important factor, although not as dominant as those mentioned above.

The density usually increases toward the bottom of the water, sometimes gradually over the entire vertical water column and sometimes

Figure 9. Boundary layer separation. See text for explanation.

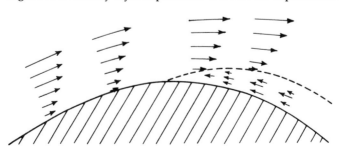

rapidly within a certain depth interval. When a strong vertical density gradient is present (common in the summertime in lakes), the turbulent velocity fluctuations are subdued or prevented and the turbulent exchange of water between contiguous layers becomes weaker or ceases. This occurs because the energy needed to move lighter layers downward and heavier layers upward increases with the density differences. Not only will the turbulent exchange be affected but also the vertical velocity distribution, the shear stress, and the transition from laminar to turbulent motion. Furthermore, if the stratification in a lake is sufficiently strong, it may even be useful to consider the lake as two bodies of water separated by the layer of strong density stratification. This definition of water masses is quite similar to the atmospheric concepts of air masses.

Whenever a stratified volume of water is present, it is possible to derive a dimensionless parameter to describe the stability of the flow. This so-called Richardson number has the form:

$$R_i = -\frac{g}{\rho}\frac{d\rho}{dz}\bigg/\left(\frac{\overline{du}}{dz}\right)^2$$

where ρ = density of the liquid, $\bar{u}$ = mean water velocity, and, g = acceleration resulting from gravity. When the liquid is homogeneous $R_i = 0$, the stratification may be called neutral. If $R_i > 0$, the stratification is considered stable; for $R_i < 0$, it is unstable.

There are some contradictory opinions of the value of R_i for the transition from laminar to turbulent flow. However, far from boundary surfaces, a value of 0.25 (Prandtl 1952) is usually accepted. Much lower values have been observed near a solid boundary.

Density currents

The effects of density stratification have been focused most predominantly on phenomena called *density currents*, in which a volume of water is flowing over, under, or through the main water mass as a result of gravity and the density characteristics of the respective water masses. In the limnological literature this phenomenon was described as early as in the classical work by Forel (1895). Although not often easy to observe directly, density currents occur frequently in lakes. For example, streams feeding into lakes very often have different density characteristics than the lakes themselves. Sometimes those density differences may even be purposely man-made to create a density current.

The persistence of a density current in a lake is mainly dependent upon the state of flow, as can be defined by the Richardson number; the density current may gradually decrease and disappear. A more rapid disappearance of the current may result from general mixing across the interface if a disturbance develops as a wavelike phenomenon similar to surface waves. If this wave travels with increasing amplitude, it finally breaks and extensive mixing occurs (Knapp 1943). Small-scale mixing occurs when the density current leaves the water surface and continues as an undercurrent or when an obstacle on the bottom disturbs the current. If the high density of a current is mainly a function of suspended load (usually caused by a high degree of turbulence), the phenomenon is called a *turbidity current*. Turbidity currents are most common in estuarine environments and in reservoirs on large rivers. They may also occur as artifically produced currents, for example, when dredging.

In an estuary, a heavy suspended load causes a very rapid current, which is characteristic for most turbidity currents. These rapid currents are likely to be erosive, as evidenced by the large canyons outside some of the world's largest river estuaries. When the velocities decrease because of the topography of the bottom of a lake or an ocean, the turbulence decreases and the suspended material settles out.

Rapidly moving turbidity currents are rare in lakes, but the slow-moving density displacements of volumes of water are common features.

Temperature properties of lakes

In addition to being the most important parameter concerning density stratification in lakes, temperature is also a fundamental factor in determining the chemical and biological processes present. The factors determining lake temperature are primarily latitude, altitude, and locality (continentality). Within low-lying equatorial areas, lakes are warm (27 °C to 30 °C) the year round with essentially no yearly, but sometimes a daily, variation. Near polar areas lake temperature is low the year round but usually with greater yearly, than the daily, variations. Some lakes are even frozen all year.

Within the midlatitudes, lake temperature is strongly variable during most of the year, and variable air temperature also creates considerable gradients in vertical lake temperature. The sun is the dominating source of energy in this respect. The radiation arriving, assuming a sufficiently large angle of incidence, has a tendency to be very quickly absorbed within the upper layers. The light in the visible range is almost totally absorbed within the upper 2 m, and of the total radiation energy entering the lake, about 50% is absorbed within a 2 m-layer depth. The radiation as such has a considerable impact on the biological processes in the lake.

The wind is then the distributing medium for the heated water within the basin. Variations in other factors – evaporation, inorganic material in the water, or long wave radiation from the water surface in the nighttime – also contribute to the distribution but only to a small extent compared to the wind. The high specific heat of the water is another important property of lake water. This property means that energy can be stored in the water, which in turn leads to the fact that water heats more slowly and cools off more slowly than the land surface. When the water cools, its density is increased. Lake water reaches its maximum density at about 4 °C. Above and below this temperature the density decreases (Figure 10). Most lakes are shallow, and to at least 100 m depth, the effect of the density caused by the water depth can be omitted for most calculations.

Figure 10. The relation between temperature and density in lake water.

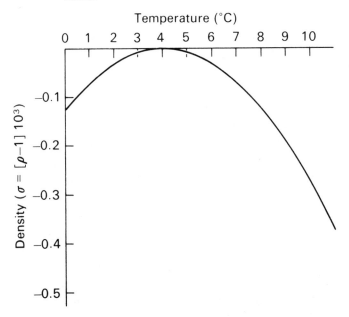

A typical midlatitude lake

Most lakes are located within the midlatitudes (glacially sculp-
tured or remodeled) and consequently undergo strong temperature
variations during the year. Thus, it will be convenient to start the
discussion about temperature in lakes with those in the midlatitudes.

Let us consider a lake with ice cover in the wintertime. When sunlight begins to have an effect in the early spring, the ice cover melts.
This process is very often accelerated in the final phases by the wind
and the rain. The actual breaking up of the ice cover is often very fast,
and usually occurs in a few hours, most often in the month of April.
At those latitudes, the spring flood often occurs at the same time and
a period of strong vertical mixing and turbulence is initiated in lakes
that have high flushing rates. When the spring flood ceases and the
radiation from the sun has increased – sometime in May – the upper
water layer begins to warm, a process that accelerates through midsummer. Until the middle of May very little energy is usually required
to mix the whole volume of water, and the water is heated to about 6–
8 °C. The length of this circulation period is dependent upon several
factors: the surface area, depth, topographic location, presence of
winds, and so on. Those factors also determine the temperature the
lake will reach before the vertical stratification starts. During a calm
and clear period, the strong summer stratification is initiated. The
density gradients rapidly increase, which means that the wind is increasingly retarded in working its mixing effect through the water
mass. The upper layer is mixed to a homogeneous temperature by
turbulence, and below this surface layer, the vertical exchange of water
is very limited, as previously discussed. Thus, usually three relatively
well-defined separate layers in the water mass are created in the summertime: an upper homogeneous layer (the *epilimnion*), a lower fairly
homogeneous cold layer (the *hypolimnion*), and an intermediate layer
characterized by a strong temperature gradient. The definition and
description of the intermediate layer has always been a matter of controversy. The most commonly used terms are thermocline and metalimnion. However, this layer has little physical or chemical uniformity.
The use of the term metalimnion, which implies a uniformity similar
to the epi- and hypolimnions, may be misleading for many readers.
Therefore, throughout this chapter the term *thermocline* is used.

A lake with a typical summer stratification has a vertical distribution
of temperature as shown in Figure 11. The late-summer lowering of the
thermocline is essentially an effect of wind. The epilimnetic water is

gradually heated until the latter part of July, and the thermocline depth increases all through the summer until it finally breaks up completely in the fall (Figure 12). The time of the start of the fall circulation (complete mixing) as well as the temperature of the water at that time is a direct function of the frequency and force of winds. The fall circulation usually starts in September. The initiation, development, and termination of the stratification is often very different from year to year (Figure 12). It can also be seen from this figure that the temperature of

Figure 11. Vertical distribution of summer temperatures in a dimictic lake, Lake Ekoln, Sweden, June 1969. (Kvarnäs and Lindell 1970)

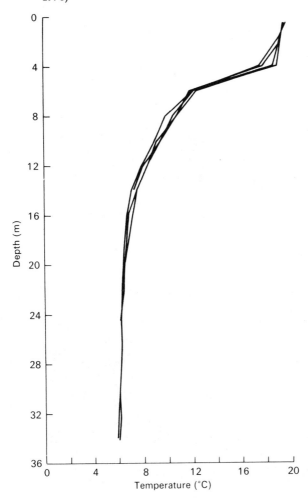

the hypolimnion gradually increases through the summer. The reason for this increase is still not satisfactorily explained. Solar radiation, turbulent transfer, and extensive mixing in times of upwelling, as well as biological decomposition, will certainly contribute heat. The concept of *upwelling* is best described as a situation in which the hypolimnetic water reaches the water surface because of strong wind effects that disturb the thermocline (see further below). Hutchinson (1957) once proposed that density currents are a main source of energy transferred to the hypolimnion (later adapted by Wetzel 1975). This process has never clearly been established and extensive sampling in Swedish lakes with the aid of very sensitive instrumentation has so far not provided clear evidence for the importance of this process. There is, however, some indication that vertical entrainment occurs in summer and may

Figure 12. Yearly distribution of temperature as isotherms, Lake Ekoln, Sweden. Dashed line is approximated. (Kvarnäs and Lindell 1970)

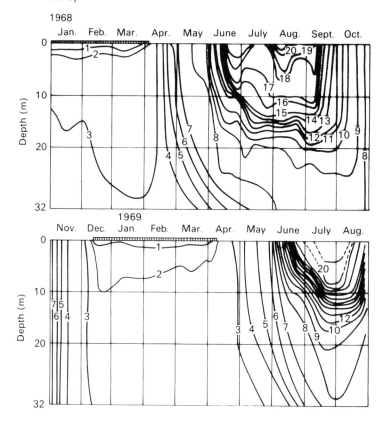

be responsible for transporting phosphorus vertically. This will be covered in Chapter 6.

In late summer the insolation decreases, and thus the income of energy to the lake decreases. The temperature decreases, and after the circulation in September, the water mass usually has a temperature of about 10 °C. The temperature then decreases under total circulation down to 4 °C or lower. As the density decreases below 4 °C, a calm, cold night (usually in December) could very rapidly cause the creation of an ice cover. The ice cover increases during the winter, and the water temperature in late March may have a vertical distribution as shown in Figure 13. The water temperature is close to 0 °C in the surface layer and close to the maximum density (temperature) in the

Figure 13. Vertical distribution of winter temperature in a dimictic lake, Lake Ekoln, Sweden, March 1970. (Kvarnäs and Lindell 1970)

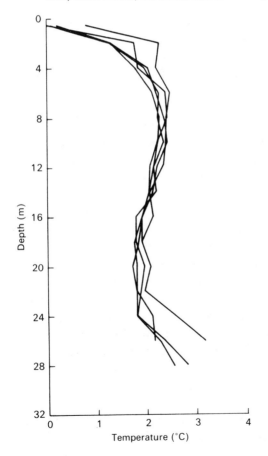

bottom layers in this case because of heating from the sediments plus inflowing water. The slight instability in the temperature distribution between 8 and 22 m is in this case compensated by a salinity gradient and is the reason the water mass is actually strongly stratified.

Classification of lakes according to temperature regimes

As the temperature conditions in lakes are very important for the physical, chemical, and biological processes, it may be fruitful to categorize the lakes according to thermal conditions in order to understand regional differences in the behavior of the lakes.

The above discussed midlatitude lake is called *dimictic*, indicating the two circulation periods. Dimictic lakes also exist at high elevations within subtropical regions. The requirement of two circulation periods is fulfilled in the midlatitudes as long as the lake is continentally located. Within maritimely imprinted areas in the midlatitudes, the winter temperature is not low enough for the lake to freeze. Thus, only one circulation period appears, and it results in a warm *monomictic* lake. The temperature during the winter is usually 4–7 °C in this type of lake. Within cold regions, a corresponding type exists, with a circulation period during the summer (temperatures up to 4 °C), which is logically called *cold monomictic*. If the climate is so extreme that the lake is always frozen, it is called *amictic*. Its opposite, a *polymictic* lake, is found within tropical areas where the yearly variation is insignificant, although short-term variations create weak stratification and periods of circulation in between. Polymixis can also occur in shallow-temperate lakes in summer, which will be shown later to greatly influence nutrient cycling. Cold polymictic lakes also exist within high-altitude tropical areas with strong winds. Within the tropics, some lakes, called *oligomictic*, circulate only sporadically. The different lake types are graphically described in Figures 14 and 15.

In the above types of lakes, temperature variation is the main cause for stratification, and circulation occurs in the whole vertical water column. A complete vertical circulation may in certain cases be suppressed by a permanent or a quasi-permanent chemical stratification of the water mass. In those rare cases lakes are termed *meromictic* and a very odd temperature stratification sometimes occurs.

Water movements

The water movement of a river is caused by gravity. The water surface usually has a substantial inclination (gradient). In a typical lake, however, the gradient current is of no importance. The gradient is considered the cause of the current in only elongated, flow-through type lakes. The most important factor in producing currents in lakes is the wind. Let us concentrate our interest on the water movements in lakes as a function of wind.

The wind causes waves on the surface of the lake from the friction against the water and also a current, which rapidly decreases with

Figure 14. Lake types classified according to temperature criteria. Dotted lines indicate winter or summer stratification. Dashed lines indicate circulation temperatures. Shaded area indicates ice cover.

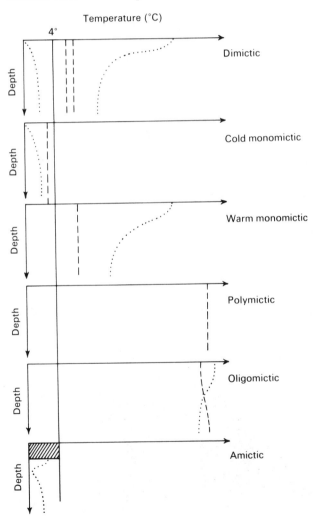

depth. By this transport of water from one side of the lake to the other an asymmetrical distribution is created in the water volume, and the continuous water transport in one direction at the surface is compensated by a transport in the opposite direction often in deeper parts of the lake. This transport of water is especially interesting in the summertime when the thermocline is inclined and the waters of the epi- and hypolimnions have practically separate circulations because of the small amount of friction between the two layers. If the wind suddenly ceases, the thermocline oscillates back and forth like a seesaw until its resting position in the horizontal plane is achieved. The rhythmic movement of the thermocline is called an *internal seiche*, and the corresponding movement on the water surface itself is called a *surface seiche*, but the oscillation of the latter is of considerably less magnitude.

In large lakes the angular velocity of the earth and the effects of such a Coriolis force must also be taken into account when dealing with water movements. In many cases current phenomena in large lakes are functions of the size of the lake and thus will not be present in small lakes.

Surface waves

Surface waves are interesting ecologically in primarily near-shore areas. In the deep parts of lakes the waves are nontransporting

Figure 15. Classification of lake types, based on latitude and altitude. Areas not labeled in the figure are transitional zones. (Wetzel 1975)

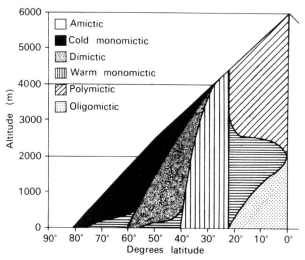

and in general each water particle can be described by an orbital movement, the circle (wave amplitude) diameter of which is halved for each $\lambda/9$ depth, where λ = wavelength. The wave height is usually about $\frac{1}{20}$ of the wavelength. The wave height is a function of the distance along the water surface in the direction of the wind (= fetch) as well as wind speed and duration. In the nearshore zone where the wave is breaking, the orbital movement is transferred to a horizontal water movement, and the wave generates an erosive character that causes the bottom substrate to be unstable and the zones affected by wave action to be highly turbid. The appearance of big waves in oceanic coastal zones or in large lakes means that the influence of the waves will be observable to greater depth; however, the velocities resulting from the waves usually die out at approximately $\lambda/2$ below the water surface. In small lakes the wave action is very limited, but within areas that are subjected to the wind, the leeward side of the lake is often shallow and heavily vegetated, whereas the windward side is deeper and biologically bare, a result of the stronger wave action on the windward side.

Wind-driven currents

The direction and velocity of wind-driven currents are determined by the classical Ekman spiral. To be able to apply this law, however, one has to assume a free water surface (no bottom and lateral friction). The Ekman current implies a surface current directed 45° to the right of the wind and a mean water transport 90° to the right of the wind because of the Coriolis force (Figure 16). In lakes, both bottom and lateral friction exists to a variable degree, and a generally accepted way of computing the currents from wind data in small lakes does not exist. Many attempts have been made based on the equations of continuity and motion, but they vary considerably in their final form.

For large lakes, the three-dimensional models for the Great Lakes by Simons (1973), based on classical oceanographic models but also incorporating the topography of the lake, have been very successful. Very simple relations have also been proposed, as for Lake Ladoga (Witting 1909), where a current velocity was calculated according to

$$v = 0.48 \sqrt{w}$$

where w = wind velocity. Witting observed a direction of the current slightly to the right of the wind. For some small Swedish lakes the following formula has been used:

$$v = c \cdot h \cdot w$$

where $c = 2 \times 10^{-5}$ (for no stratification h = mean depth) and $c = 1.5 \times 10^{-5}$ (for h = mean depth of the thermocline). As a rough estimate of surface currents in lakes, it is possible to consider them parallel to the wind and on the order of 2% of the wind velocity. In the deeper parts of lakes the water movements are slow, generally only a few centimeters per second, which presents severe problems in recording water movements because the velocities are lower than the activating velocity of most current meters.

Seiches

When the wind blows, it causes a piling up of the water on the windward side of the lake. The inclination of the water surface resulting from wind is usually on the order of a few decimeters in large lakes and a few centimeters in small lakes. Morphometric conditions in lakes are, however, strongly modifying, and consequently, general rules are difficult to postulate. For a rectangular-shaped lake, Hellström (1941) suggested the inclination to be given by

$$s = \frac{3.2 \times 10^{-6}}{g\bar{z}} \cdot w^2 l$$

where s is expressed as the total difference in height between the windward side and the leeside of the lake, g = acceleration resulting

Figure 16. The distribution of water movements in a wind-driven ocean (the Ekman current). (Hellström 1940)

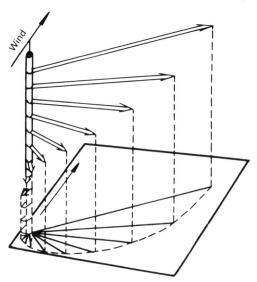

from gravity, $\bar{z}$ = mean depth, w = wind velocity, and l = length of the basin.

The oscillation of the water surface after the wind has ceased may then be determined according to the classical formula by Chrystal (1904):

$$T_n = \frac{l}{n} \frac{2l}{\sqrt{g\bar{z}}}$$

where the ratio of the periodicities for the uninodal, binodal, and trinodal seiches are $T_1:T_2:T_3 = 100:50:33$ and the nodes at $x = 0$ (uninodal), $x = \pm 0.25l$ (binodal), and $x = \pm 0.33l$ (trinodal); and l = length of the basin, g = acceleration resulting from gravity, $\bar{z}$ = mean depth, and n = number of nodes.

This general formula is derived from a rectangular basin with constant depth where the seiche can be considered a shallow water wave, but it has been shown to be applicable to most types of lakes. Surface seiches generally have a fairly small impact on the variability of water-quality characteristics. The internal seiche could, however, cause strong fluctuations in water quality both in time and space because it is separating two water masses that could be of very different quality. This seiche has an amplitude of a magnitude 100–1000 times larger than the surface seiche and is primarily determined by the density differences between the epilimnion and hypolimnion. The periodic motion of the internal seiche can be determined (Mortimer 1952) by:

$$S_n = \frac{2l}{\left[g(\rho_h - \rho_e) \middle/ \left(\frac{\rho_h}{h_h} + \frac{\rho_e}{h_e} \right) \right]^{0.5}}$$

where h_e = thickness of epilimnion, h_h = thickness of hypolimnion, ρ = density, and l = length of the basin.

The period of the internal seiche is often on the order of hours, whereas the surface seiche is calculated in minutes. Because it is sometimes difficult to record the slow water movements in lakes, as mentioned previously, the mean velocities associated with the internal seiche can be roughly estimated (Mortimer 1952) as:

$$V_e = \frac{sl^2}{2Th_e}$$

and

$$V_h = \frac{sl^2}{2Th_h}$$

where s = maximum slope of the thermocline, l = length of the basin, T = period, and h_e, h_h = depth of epilimnion and hypolimnion respectively.

Tidal effects

The tide is mainly a function of the gravitation of the moon. The effect of the sun's gravitation can be observed only in connection with extreme high or low tide. Tides are probably insignificant in the explanation of water-quality changes in lakes. Very little knowledge has been generated so far in this field, but it is believed that tides may cause waves in the order of a few centimeters in large lakes.

In oceanic coastal zones the tides are highly variable because of the general distribution of the land masses and the local morphometry of the coastal zone. In certain areas the tides could thus be close to zero, whereas in other places, especially in funnel-shaped estuaries, the tides may be of the order of several meters. The tides have two maxima and two minima per day for a lunar day that is 24 hours 50 minutes. An estuary with strong variations between high and low tide, therefore, has an extraordinarily good transport mechanism, as flushing occurs twice a day.

Water movements in large lakes

In small lakes the effect of the earth's rotation may be detectable in the estuary of some inflowing water but hardly detectable in the internal water movements. In large lakes, however, the Coriolis force is observable on all water movements and in very large lakes the Coriolis force may totally dominate the current patterns. The meaning of "large" as used here must be defined in relation to the latitude, and the Coriolis parameter, f, must be introduced:

$$f = 2 \, \Omega \sin \phi$$

where Ω = angular velocity of the earth and ϕ = latitude.

The Coriolis force is directed perpendicular to the right of the motion in the Northern Hemisphere and to the left in the Southern Hemisphere. As can be judged from the definition, the effect of the Coriolis force increases with latitude and is zero at the equator.

By using the definition of the Coriolis parameter, it is now possible to calculate the radius of the inertia circle:

$$r = \frac{u}{f}$$

where u = water velocity.

For a typical current velocity of 10 cm per sec at 45° latitude, for example, $r \sim 1$ km. If the width of the lake is 5 times the radius, the effects of the earth's rotation can be observed, and if the width is 20 times the radius or more, the Coriolis force will dominate the currents.

Kelvin waves

If a seiche is dominated by the Coriolis force in a sufficiently large basin, it could result in a particular motion called a Kelvin wave. Instead of the creation of a standing wave in a small lake, a Kelvin wave progresses counterclockwise (Northern Hemisphere) around the basin trapped by the coastline. Its maximum altitude occurs close to the coastline and decreases very rapidly toward the center of the lake. In still larger lakes a fairly complex wave pattern may occur with areas of alternating higher and lower levels and with isolated circulation within the cellular structures. Sometimes combinations of Kelvin and this so-called Poincaré-type wave may occur.

A very typical feature of large lakes is rapid water movement in nearshore areas. It is now an accepted fact that the "coastal jets" are sometimes connected to a type of deformed Kelvin wave, resulting in a very asymmetrical wave that progresses counterclockwise with a very rapid countercurrent (clockwise) on its rear side.

The coastal jets are otherwise very common in large lakes. When a wind initiates Ekman drift in the center of a lake, the Coriolis force balances the wind stress. In a narrow coastal band, however, the wind stress accelerates the water alongshore. This mechanism transports water to the windward side of the lake, and when the wind ceases, a return flow develops that is again confined to the nearshore bands.

In the summertime on the lakeward side of coastal jets a persistent geostrophic counterclockwise circulation dominates (Figure 17), caused by a combination of wind, Coriolis force, and temperature contrasts. This circulation is often so strong that the common and substantial vertical shifting of the thermocline resulting from wind never overcomes the effect of the circulation.

Thermal bar

A certain type of circulation common in large dimictic lakes in the springtime (and theoretically also in the fall), caused by the uneven heating of the water, is called *thermal bar*. This phenomenon can be illustrated using the spring situation.

If the temperature of the whole water volume of the lake is well below 4 °C and the heating during the spring is rapid, the rate of heating is greatest in the nearshore shallow areas. The deep pelagic areas of the lake remain below or near 4 °C. In the transitional zone between the high temperatures of the coastal areas and the cold open lake areas there is a zone with a strong horizontal temperature gradient (Figure 18). This zone moves slowly from the nearshore areas toward the open waters. The duration of this process varies from days to months, depending on the differences in heating and size of the lake. The typical thermal-bar phenomenon is transformed to the temperature pattern that is typical for geostrophic water movements. The currents associated with these water movement phenomena are characterized by a predominant alongshore counterclockwise circulation in the nearshore stratified zone. Modeling of the thermal bar has also

Figure 17. Geostrophic circulation of a large lake (Lake Vänern, Sweden). Along a cross section (A) of the lake the temperature pattern is dome-shaped (C). The resulting dynamic pattern of height anomalies (in centimeters) is shown on B. (Lindell 1975)

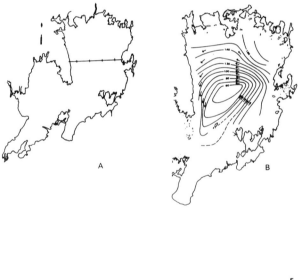

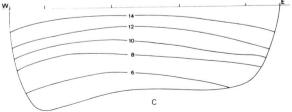

shown downward movement on the lakeside of the bar and upwelling in the warm water. The development from the early spring situation to the summer circulation is not yet satisfactorily explained theoretically.

The thermal bar is interesting because the nearshore warm waters are trapped close to the coastline, which in turn could cause high nutrient levels and algal blooms. Such a nearshore effect has been suggested in the Great Lakes (Beeton and Edmondson 1972).

Dispersal processes

In principle it is possible to separate the dispersion processes that occur in small-scale and large-scale phenomena. In lakes the small-scale diffusion processes appear in the neighborhood of the sources of pollutants or water courses, whereas the large-scale diffusion processes are closely associated with the pelagic areas of the lake. The parameters that influence the dispersion processes are water velocity and eddy diffusivity in the horizontal (K_y) and vertical (K_z) direction. K_y and K_z are dependent upon the character of the turbulence, which in turn is dependent upon water velocity and stability, as has been mentioned earlier. They are also dependent on the scale of the phenomenon.

Figure 18. The temperature (°C) distribution in Lake Ontario (surface and longitudinal section) during June 7–10, 1965, showing the position of a thermal bar. (Rodgers 1966)

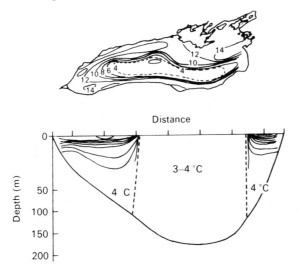

Horizontal diffusion
The horizontal eddy diffusion along the horizontal plane (*x*) may be defined as

$$K_y = k \cdot u \frac{dw^2}{dx}$$

where u = water velocity, in the x direction; w = width of the plume, in the y direction; and k = constant.

Csanady (1970) has proposed a value of .03 for the constant in the equation. The diffusivity increases rapidly with the size of the plume. The increase in the width of the plume exceeds that of a simple linear increase. Although resulting from turbulence, the dispersion process at sufficient distances from the source becomes more similar to molecular diffusion. This development is the consequence of the growing plume, causing an increase in the size of turbulent eddies, which become increasingly important. The size of the turbulent eddies does have an upper limit, which is why this type of diffusion eventually becomes similar to a molecular diffusion. Although the processes are nonlinear with complicated interrelationships, a linear diffusion model is often sufficiently good for most purposes. A typical development of a plume from a dye injection will be related to distance as shown in Figure 19. Csanady (1970) defines plume diameter as about 10% width (*w*) of the cross section at the surface. The horizontal diffusion coefficient is usually in the order of 100–1000 cm² per sec in lakes.

Figure 19. Horizontal diffusion. Changes of the width of a plume in the direction of the length axis.

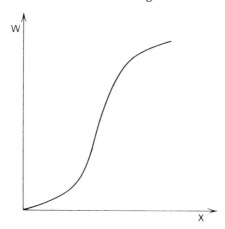

Vertical diffusion

Vertical diffusion has certain similarities with horizontal diffusion, but the magnitude of vertical diffusion can be affected by many important variables, such as density, water velocity, and current direction, which can change rapidly with depth. Vertical diffusion is especially different in the presence of a thermocline. As noted earlier in discussing density, it is sometimes useful to consider a lake as two lakes with the epilimnion as the "upper lake." The lower limit of the epilimnion acts as a floor in the process of diffusion. The two main types of vertical diffusion may, therefore, be generalized according to Figure 20. Typical vertical diffusion values are $\sim$ 5 cm^2 per sec, but initially much higher values (up to 30 cm^2 per sec) are common (Csanady 1970). Compared to horizontal diffusion, vertical diffusion is much lower because of the differences in the horizontal and vertical structures of turbulence.

Large-scale phenomena and conditions in large lakes

The most important large-scale diffusion process for most sizes of lakes is the meandering type that follows a large inflow. Meander diffusion is similar in appearance to the meandering of rivers; that is,

Figure 20. Vertical diffusion. Typical patterns of diffusion in Lake Huron in the absence (A) and presence (B) of a thermocline (from Csanady 1970)

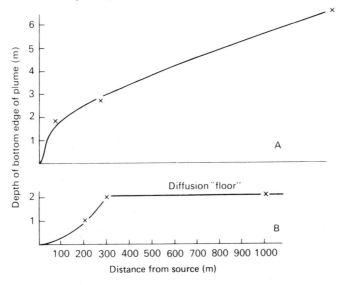

the movement is sinuosidal in plain view with a helicoidal component. It is not possible to give any general rules for the occurrence and frequency of meandering in lakes.

In large lakes where thermal bars and coastal jets exist, the dispersion of pollutants often leads to concentrated levels in the coastal zones, thus restricting horizontal diffusion. It should be noted, however, that the jets and thermal bars vary substantially with time, and the diffusivities, therefore, could be very different from one day to the next.

Typical dispersion patterns of plumes
The following types of dispersion may occur:
1. Complete mixing
2. The pollutant usually disperses horizontally
   a. At the surface
   b. At an intermediate level
   c. Along the bottom
3. Combinations of 1 and 2

Complete mixing is almost entirely a function of the velocity of the receiving water because diffusion coefficients depend on velocity. A rapidly flowing stream always has a high degree of turbulence, and, consequently, an input from a point source or a nonpoint source can be homogeneously distributed within the whole river width usually after only a few hundred meters. However, the distance will depend on how the pollutant enters the stream and the stream size. It is easy to be misled by this fact when considering the water quality of a river system. To determine the water quality of a downstream lake, in addition to observing pollutant distribution near the input, the total transport per unit time of a contaminant must be calculated. To calculate the total transport of a contaminant requires knowledge of the quantity of water as well as the average concentration. In a strongly turbulent stream there is little interest in whether the pollutant is dispersed as a concentrated jet or by the diffusion process. The turbulence of the stream immediately dominates the dispersion of material, and the resulting cross-sectional concentration may become homogeneous a few hundred meters downstream if the stream is of moderate width. Consequently, adverse effects are noted in turbulent waters downstream from the mixing zone only at relatively higher pollutant-loading rates, whereas in a nonturbulent stream adverse effects (or "hot spots") are more easily noted.

If the densities of the receiving water and a pollutant are considerably different and the water velocity is lower, the pollutant mixes into the stream very slowly. High concentrations of the discharged pollutant often remain unmixed very far downstream – sometimes several kilometers – if the degree of turbulence is low, the density gradients are strong, and the stream width is substantial. Considerable variability exists in the mixing processes because of density structures and bottom topography. The most typical form of dispersion occurs, however, when the usually heavily loaded incoming water progresses close to the bottom. When the content of dissolved solids makes the incoming water extremely dense, the temperature has little effect on density and hence dispersion.

When the density differences are considerably less, the situation is equally critical. The incoming water mass tends to stick to the entering side of the river for a long distance, still assuming that fairly low river water velocities exist. This process is easy to observe naturally, for instance, at the confluence of two tributaries carrying very different loads of suspended sediment.

In streams with this type of incomplete discharge mixing the ecological disturbances may be severe on one side of the river and nonexistent on the other. The monitoring of stream water quality seldom includes samples from the entire river width. An incomplete mixing could, therefore, easily lead to nonrepresentative water-quality sampling, and local samples cannot be considered to give spatially averaged concentrations, which would occur if the pollutant was uniformly distributed over the cross section.

A wide variety of different transformations that are mainly a function of the water velocity but also to some extent of the density patterns also occur. To accurately monitor stream water quality usually requires a careful study of the turbulence patterns during different conditions of stream flow, for example, with the aid of tracer studies.

Although dispersion and mixing in estuaries and lakes are restricted by large density differences, the mixing processes, on the other hand, are accelerated by wind and currents in the open waters. The diffusion process in a lake or an estuary is fairly simple to define from a number of theoretical linear models, but the complete modeling and forecasting of the development of a plume is very complicated, not only because of the complexity of the model itself, but also because of the stochastic nature of the varying meteorological and hydrological conditions. However, a few descriptive cases of mixing will be discussed.

An estuary from a stream discharging into a lake is similar to the point-source input to a stream. The main types of mixing are (1) complete mixing and (2) mixing that is density dependent at the surface, bottom, or intermediate inflow level.

The basic condition governing complete mixing is a nonstratified water mass (spring and fall for a dimictic lake). This is most often an adequate assumption because the density of the stream seldom deviates so much from the lake that it inhibits mixing. Only very dense incoming water, close to the density maximum at 4 °C plus a dense load of suspended or dissolved material, dives as a density current toward the bottom of the lake and remains there for some time. The best conditions to obtain a low concentration of a pollutant in lakes occur during the normal spring-fall mixing periods. During the summer stratification, essentially three types of dispersion and mixing occur. The inflow of water stays at the surface of the lake if the density of the incoming water is lower than that of the lake water. The process is most simply illustrated in Figure 21.

The early summer period, when this type of flow condition first occurs, is usually the most suitable time to observe it. During this part of the year the incoming water is warmer than the lake water (unless the river water is meltwater) because the landmass warms up more rapidly than the water and the volume of incoming water is usually small relative to the lake volume. This phase often starts at the end of the spring flood in dimictic lakes and ends as the incoming water begins to dive later in the summer. In the beginning of this phase the initial mixing is quite strong because of the high velocity of the incoming water. The water of the streams gradually becomes more concentrated as evaporation processes are initiated. The velocities of the streams also gradually decrease and the inflowing water dives to a typical position at the upper level of the thermocline. This inflow at the top of the thermocline is generally very distinct with little initial

Figure 21. Surface inflow in a lake.

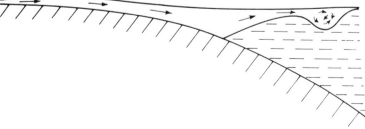

mixing because of the low water velocities. The dispersion of the inflowing water in the lake is usually in horizontal and longitudinal directions. Furthermore, it is possible in some lakes to follow an inflowing stream throughout the whole lake to the outlet (Figure 22).

The remaining type of inflow, which dives to the near-bottom layers, usually exists in the wintertime. This type is analogous to the bottom inflow in rivers as described earlier. Occasionally, when little temperature differences exist between the lake and the inflowing water, only a small increase in dissolved substances is sufficient to create this bottom flow. Because river water during the winter is slow moving, the initial mixing in dimictic lakes is small. The ice cover also protects the lake from direct wind effects. It is a common feature in dimictic lakes in winter for inflowing water to accumulate in the deeper parts of the lake and not mix with the rest of the lake water until the spring flood occurs (Figure 23).

In low-elevation estuaries there is sometimes an interaction between the water masses of the sea or the lake and the incoming water. The initial mixing zone could vary greatly because of differences in tides, discharges, and wind effects. Lake water, for example, may be forced far upstream of the estuary under a strong onshore wind, and the initial mixing thus occurs not at the river mouth but several hundred meters upstream (Figure 24).

In an estuary facing the sea, low-density inflowing water flows on top of the salt water, and the initial mixing zone, as well as the degree of mixing, is dependent primarily upon the ratio of tidal prism volume and volume of river inflow per tidal cycle. An estuary tends to be more highly stratified when the river inflow is large compared to the tidal prism, and hence mixing associated with tidal energy is not so large. In very exposed shallow estuaries, however, the wind is as important a factor as the tide.

Study questions
1. What types of water flow exist and why does just one type dominate in nature?
2. What regimes of flow exist and what are the characteristics of each type?
3. Discuss the causes for density gradients in rivers, lakes, and estuaries.
4. How does a density gradient influence the vertical mixing in a lake?

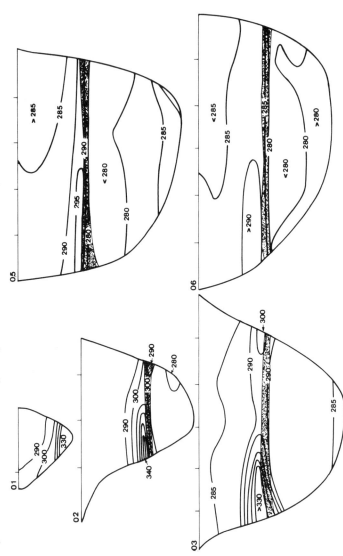

Figure 22. Cross-sectional recordings of specific conductance in the River Fyris within Lake Ekoln, Sweden, August 1970.

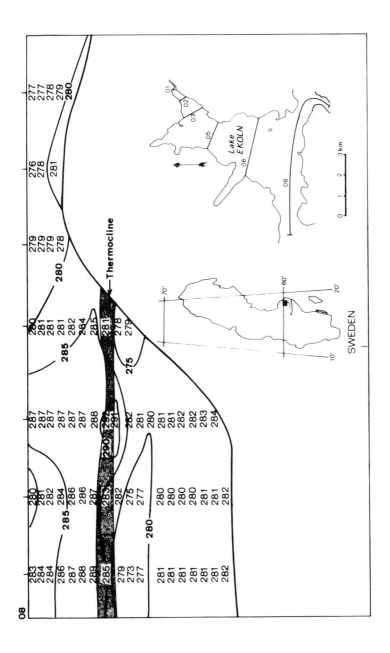

Figure 23. Bottom inflow in a lake.

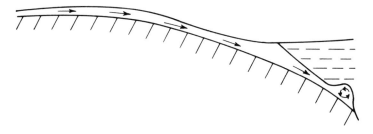

Figure 24. Interaction between river water and lake water at different flow conditions and different wind directions. Open arrows indicate wind, solid arrows indicate flow of water. The river water is shaded.

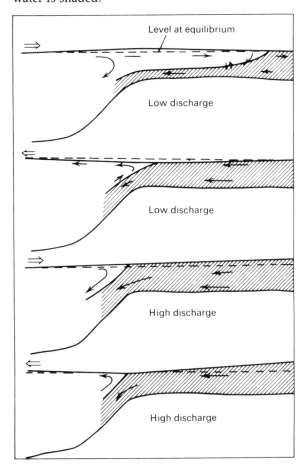

5. What does the Richardson number mean?
6. What are the characteristics of density and turbidity currents?
7. Describe and explain the temperature regime of a midlatitude lake.
8. Mention two other types of temperature regimes in lakes?
9. What types of seiches exist?
10. What mechanism creates an internal seiche?
11. Calculate the uninodal surface seiche and internal seiche in a rectangluar basin 2 km long with a mean depth of 15 m containing an epilimnetic layer of 5 m and a hypolimnetic one of 10 m and with mean temperatures of 18 °C and 9 °C respectively.
12. Calculate the radius of the inertia circle of a particular lake and discuss how that figure affects the water movements of the lake.
13. Define thermal bar and describe its characteristics.
14. What are the principal differences between horizontal and vertical diffusion?
15. Discuss the different types of dispersion patterns that may occur in a lake.

4

POPULATION GROWTH

Unlimited environment

Growth of any population of living organisms can be described as exponential in an unlimited environment. Individuals (or biomass) are added at an increasing rate as the density of the population becomes larger. In that way population growth is similar to compound interest – the interest, or biomass, increase being recalculated at each time increment on the new balance, or biomass, attained at time t. Thus, the rate of population increase depends not only upon its growth rate but also on its initial size (see Kormondy 1969). The following equation defines unlimited growth:

$$\frac{dN}{dt} = rN \tag{1}$$

where r is the growth rate constant and N is population size (Figure 25).

Limited environment

In reality, population density cannot continue to increase exponentially because resources are not inexhaustible; the resource exhausted first slows and ultimately stops the increase. Populations that grow rapidly, seemingly to "explode," such as spring plankton blooms, tend to exhaust the limiting resource and reach their maximum attainable biomass all at once, which is followed by a crash – such growth curves are J shaped. If environmental resistance is gradual, the growth curve takes on more of an S shape, passing through such phases as lag, accelerated growth, maximum growth, decelerated growth, no growth, and death (Odum 1959). Of course, during the phase of maximum growth rate, which is the steepest slope of the curve, the rate of population increase is also greatest (Figure 26).

The general equation that defines population growth that is environmentally limited is:

$$\frac{dN}{dt} = rN \left(\frac{K - N}{K} \right) \tag{2}$$

where K is the maximum attainable population size and r and N are as defined previously.

Unicellular vs. multicellular organism growth

The growth rate constant, r, has a maximum for each population that is consistent with its genetic capacity in an optimum environment. If the environment is less than optimum, r will not reach the genetically fixed maximum. In multicellular, sexually reproducing organisms, r is affected by the age structure, or that fraction of the population that is reproductively active. If the age structure remains constant, r can be evaluated with respect to environmental variables. However, much of the change in r could be the result of altered age structure. This is not a problem in the analysis of unicellular, asexual populations because all have an equal chance at reproduction. In both types of populations, r = birth rate − death rate. Typical r values in nature can range from 0.01 per year for the U.S. human population to as much as 2.0 per day for natural populations of blue-green algae and even on the order of 3.0 per hour for bacteria.

Figure 25. An unlimited population growth curve. Note that population increment b is much greater than a for an equal-time increment, which is due to the larger population size. This is a result of the "compound interest" phenomenon characteristic of populations.

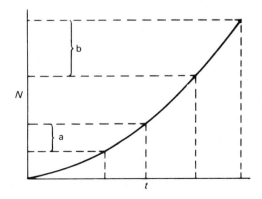

Microorganism growth and environment

The symbol describing the rate constant for microorganism growth is μ and it varies with environmertal factors in particular ways. If a limiting nutrient is considered using the Michaelis-Menten relationship developed from enzyme kinetics, bacterial growth can be described in relation to a limiting organic substrate and algal growth in relation to a limiting inorganic nutrient (Herbert et al. 1956; Droop 1973). The equation for microorganism growth limited by a single nutrient is:

$$\mu = \mu_m \left(\frac{N}{K_N + N} \right) \tag{3}$$

where μ_m is the maximum rate attainable for that population, N is the concentration of limiting nutrient, or if in substrate concentrations is indicated by S, and K_N is the nutrient concentration at one-half μ_m (sometimes referred to as K_S). This relationship is illustrated in Figure 27 along with a visual definition of the parameters. If the Michaelis-Menten relation is substituted into the first order growth equation $dX/dt = \mu X$, where X indicates biomass or population size, then

$$\frac{dX}{dt} = \mu_m X \left(\frac{N}{K_N + N} \right) \tag{4}$$

Figure 26. Population growth curve, environmentally limited, showing changes in size and rate of increase. Growth curve phases are A, lag; B, accelerated growth; C, maximum rate; D, decelerated growth; E, stationary; and F, death.

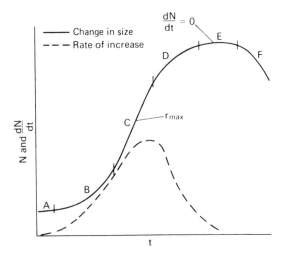

This equation then describes the biomass change in response to one limiting nutrient. If the population is cultured in a mixed continuous flow system, the dilution rate, D (inflow ÷ vessel volume), should be included as a biomass loss-rate term. The equation will then take the form

$$\frac{dX}{dt} = \mu_m X \left(\frac{N}{K_N + N} \right) - DX \tag{5}$$

Steady-state system

This equation can be applied to continuous-culture experiments in which a steady-state biomass can be maintained in a well-mixed growth chamber when cell growth, $\mu_m X[N/(K_N + N)]$, equals cell washout, DX. The dilution rate must be less than the cell growth rate for the chamber biomass to attain a steady state. If D is greater than cell growth rate, the loss from the system will be greater than the gains and the population will diminish.

To describe this system further, consider the relationships of three variables with dilution rate, cell biomass, cell output or production, and nutrient concentration in the chamber. These hypothetical relationships are shown in Figure 28 (see Herbert et al. 1956 for further explanation). Some significant points about microorganism growth that can be illustrated are:

1. The nutrient concentration that is detectable in the mixed chamber is independent of the inflow concentration but is controlled by the dilution rate, because cells are actively taking up the nutrient and quickly depleting it down to a very low level. This in itself is an indication that the nutrient that is

Figure 27. A hypothetical relationship between microorganism growth rate and the concentration of limiting nutrient.

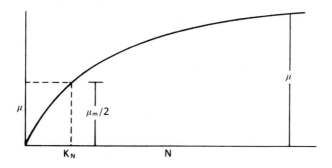

lowest in concentration, relative to the inflow concentration, is the limiting one.

2. Dilution rate determines output or production of cells. As dilution rate increases and nutrient concentration in the inflow stays constant, the nutrient supply rate increases in proportion to the dilution rate. As the nutrient is quickly and efficiently removed to a low level in the chamber, and is thus transformed into cells, the increased quantity of cell mass formed is washed out, raising the cell output rate or production.

3. Organism concentration or biomass is largely determined by the inflow nutrient concentration. If the dilution rate is held constant and the nutrient concentration in the inflow is increased, the additional nutrient put into the system goes into biomass because the loss rate, which is proportional to dilution rate, is also constant.

In summary, an increase in nutrient concentration in the inflow increases biomass and production if that nutrient is limiting, whereas an increase in dilution rate only increases production.

Such considerations of the effects of nutrient concentration, nutrient supply rate, and dilution rate are applicable to small lakes with large watersheds or large volume tributary streams, constricted estuaries, and sewage treatment lagoons. These are ecosystems with retention times on the order of 10 days or less, dilution rates of 0.1 per day or greater.

Non-steady-state system

Population growth in ecosystems with longer detention times is more typical of growth in a closed container, that is, a batch culture.

Figure 28. Dilution rate versus biomass, production, and nutrient concentration in a chemostat culture. (Herbert et al. 1956)

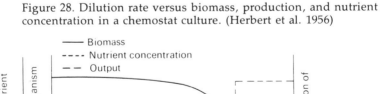
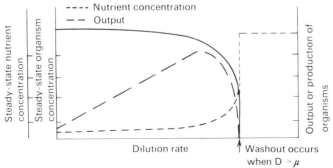

In this system the equations describing population growth, size, and nutrient concentration still apply, but a steady state is not reached. This non-steady-state condition is typical of natural systems where cell biomass and nutrient concentration continually change. A typical pattern of growth in a batch culture is shown in Figure 29. Nutrient concentration decreases as biomass increases, having been transferred into cells. The rate of biomass increase begins to slow following the period of maximum growth rate, which occurs when the nutrient content reaches a growth-limiting level.

A more complete equation to describe algal biomass change or production has two additional loss terms (Uhlmann 1971),

$$\frac{dX}{dt} = \mu X - DX - GX - SX \tag{6}$$

where, besides dilution rate (D), the additional loss rates are grazing rate (G) and sinking rate (S). Of course, growth rate (μ) is dependent upon light and nutrient concentration. The integrated form of the equation is:

$$X_t = X_o e^{(\mu - D - G - S)t} \tag{7}$$

To illustrate the significance of loss rates in controlling algal biomass three different rates for combined loss have been calculated and plotted in Figure 30. In this case μ was 1.0 per day, G and S were held constant at 0.5 per day and 0.2 per day, respectively, and D was tested at 0.05, 0.2, and 0.4 per day. Thus, the biomass change rates were +0.25, +0.1,

Figure 29. Nutrient concentration and biomass in an algal batch culture showing initial and final levels. (After McGauhey et al. 1968, p. 8)

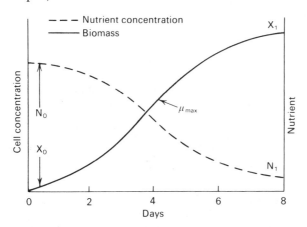

and -0.1. Clearly, biomass is very sensitive to the magnitude of the loss rates.

Use of a steady-state concept

The effect of nutrient concentration and supply rate on natural plankton populations can be hypothesized from steady-state considerations. Even though the natural system seldom reaches steady state, its tendency is in that direction. If the mechanisms that produce steady state in cultures are operating, two facts become important in understanding natural systems: (1) growth rate and biomass of plankton algae are determined by the concentration of nutrient and (2) the ultimate production is determined by the nutrient supply rate, which includes remineralization and recycling within the system as well as inflow.

Figure 30. Biomass rate of change in a hypothetical algal population with $\mu = 1.0$ day^{-1} and combined loss rates of 0.75, 0.9, and 1.1 day^{-1}.

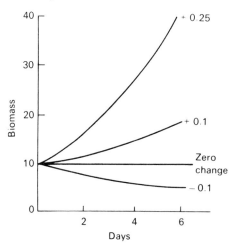

Figure 31. Idealized steady-state system for plankton limited by nitrogen. (After Dugdale 1967)

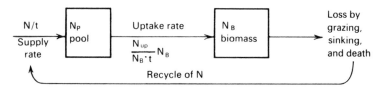

For example, consider a steady-state system in which nitrogen (N) is the limiting factor (Figure 31). Some implications here are that production (or loss) could theoretically remain high even if the nutrient pool size (N_P) is reduced by the spring plankton outburst. This could result in a reduced pool concentration (N_P), which subsequently would reduce the uptake rate, but because the biomass is greater as a result of the outburst, it tends to compensate for the reduced uptake, thus giving a similar steady-state production; production of $N_B = \ \downarrow [N_{up}/(N_B \cdot t)] \cdot N_B \uparrow$. The nutrient supply rate would remain relatively high because the increase in loss rates would supply more regenerated nutrient. However, biomass would eventually begin to decline because recycling is not 100% efficient.

A supply-rate increase of the limiting nutrient to actively growing populations results in an increased production and possibly an increased biomass (if loss rates remain low) without a noticeable increase occurring in nutrient pool size. If loss rates also increase, little increase in biomass results. Such dynamics are usually involved and can provide a reasonable explanation to typically occurring phenomena in natural systems. Although considerable management and mismanagement of aquatic systems has occurred through planned and unplanned overfertilization, little use of loss rates has been attempted. Loss rates can have as much control over biomass as nutrient supply, as illustrated in the steady-state culture system.

Study questions

1. Two alternatives exist for improving the visual quality of Lake Scummy: (a) increase the dilution (flushing) rate, and (b) decrease the concentration of limiting nutrient (phosphorus) in the inflow. The present detention time is 1 yr and the average inflow P concentration is 100 μg l^{-1}. Based on your knowledge of continuous-flow algal cultures, which alternative would you choose and why?

2. An increased dilution (flushing) rate will lead to decreased biomass if the
   a. limiting nutrient concentration in the inflow is not changed
   b. limiting nutrient concentration in the reactor has not been utilized to low concentrations
   c. growth rate does not change
   d. light intensity remains constant

3. Net primary productivity is
   a. usually equal to respiration
   b. the same as net increase in biomass in excess of other losses (e.g., grazing, sedimentation, etc.)
   c. the same as net increase in biomass in the absence of other losses (e.g., grazing, sedimentation, etc.)
   d. usually greater than respiration
4. The production of cells in a steady-state system can be increased to a new steady state by
   a. increasing the dilution rate and therefore the nutrient supply rate
   b. increasing the dilution rate but not to exceed the growth rate
   c. increasing the inflow concentration of nutrient or the dilution rate but not to exceed the growth rate
   d. decreasing the loss rate from the system by returning a portion of the outflow to the growth chamber

5

LIMITATION, TOLERANCE, AND ADAPTATION

The purpose in this section is to give a brief introduction to some ecological concepts upon which later interpretations about waste effects on populations are based. For more detail the reader is encouraged to see Odum (1959) or Kormondy (1969).

Limitation

When the question arises as to which nutrient(s) is limiting in a particular environment, the first concept considered is that of Justus Liebig, the law of the minimum, which states that the yield of a crop will be limited by the essential nutrient that is most scarce in the environment relative to the needs of the organism. "Relative to the organism's needs" is the key to understanding and using this concept. Just because an element is lowest in concentration does not mean it is the critical limiter. For example, nutrient A may be twice as abundant in the environment as nutrients B and C, but if the production of plant mass requires three times the amount of nutrient A than nutrients B and C, insufficient A will limit further production. Limitation can also occur from scarcity of more than one nutrient simultaneously if all are reduced proportionately but still remain in the required ratio for plant growth.

The yield limitation of a crop of microorganisms by the scarcity of a critical nutrient is illustrated in Figure 32. Here the maximum growth rate can vary with the temperature, light intensity, nutrient concentration, or species in the test, but the maximum yield attained depends on the same "minimum" nutrient. This concept is the basis for the algal growth potential (AGP) measurement now frequently used in the study of eutrophication and patterned after Skulberg (1965).

Another important concept is that the maximum yield (or biomass) attained in a series of AGP tests is linearly related to the initial concentration (or supply) of the most limiting nutrient. This is illustrated in Figure 33. The supply of the critical nutrient is indicated as the independent variable because, as will be shown later, the existing concentration is not always a good indicator of the total availability of a nutrient for biomass formation.

While the law of the minimum solves many problems of nutrient limitation, multiple nutrient limitation can still occur even if the final yield is determined by one nutrient. However, the multiple limitation concept is consistent with the one-nutrient-at-a-time idea because multiple limitation occurs on growth or uptake rate and not yield. The multiple limitation concept is reasonable because with every required nutrient a given species has a characteristic uptake or growth rate curve. For every combination of nutrient concentrations, even though the same nutrient would limit yield, there should be a total "resultant" growth or uptake rate. This can be illustrated by the multiplicative model based on the Michaelis-Menten principle (Chen 1970),

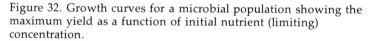

$$\mu = \mu_{max} \cdot \frac{P}{K_P + P} \cdot \frac{N}{K_N + N} \cdot \frac{L}{K_L + L}$$

where P, N, and L are phosphorus, nitrogen, and light intensity, respectively.

Figure 32. Growth curves for a microbial population showing the maximum yield as a function of initial nutrient (limiting) concentration.

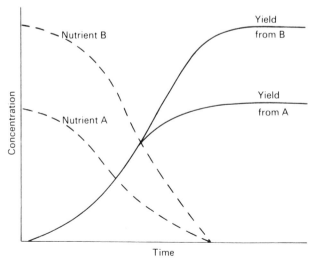

Although this model has been rather popular, there have been other approaches to model growth or uptake rate. Of particular interest is one that allows limitation by only one nutrient at a time (Bierman et al. 1973).

Because growth or uptake rate can be limited by more or even different nutrients at any one time than would limit maximum yield, the correct community characteristic being limited (growth or uptake rate or maximum biomass [yield]) must be precisely stated. Depending upon the question asked about nutrient limitation, the answer could well be different.

To illustrate the concept of limitation further, the availability of a limiting nutrient can be compared to quantity within plankton. Plankton (zoo- and phyto-) in marine waters are composed primarily of C, H, O, N, and P and on the average in the ratio of $O:C:N:P$ (Sverdrup et al. 1942):

      212:106:16:1   by atoms
      109:41:7.2:1   by weight

Although N and P make up less than the other two elements in living material, they are far more scarce in natural waters, and thus as countless studies have indicated, either N or P or both are usually the nutrients that limit growth (biomass) of phytoplankton (the necessary element[s] present in amounts nearest the critical minimum). As a general rule, C limits growth only in waters heavily fertilized with P and N. Trace elements have been observed to limit in a few instances. Although C and trace-element deficiency may temporarily limit growth rate, N and/or P deficiency usually limits growth rate and/or further increases in biomass, or yield. The role of C and trace elements may be greater in species succession, as will be discussed later.

Figure 33. Maximum yield of a microbial population as a function of the supply of limiting nutrient.

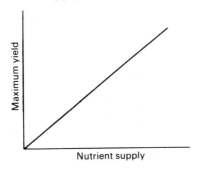

The ratio of C:N:P in biomass is often used to indicate which nutrient may be limiting. In seawater N is usually considered to be limiting. Atomic N:P ratios in N-limited phytoplankton populations in the sea were found to range from 5.4-17, whereas the ratio of C:P ranged from 67-91 (Menzel and Ryther 1964). Thus P is available in apparently adequate amounts, N is most scarce, and C is intermediate in availability.

In fresh water, on the other hand, P is usually the limiting factor. For example, the particulate matter (mostly phytoplankton) in Canadarago Lake, New York, showed a range in N:P during summer stratification of 17-24, well above 16, and C:P was about 200, which indicates that P was limiting. Soluble P content in the epilimnion during summer was undetectable (Fuhs et al. 1972).

On a biospheric average in fresh water, the plant demand, determined by the average concentration of the given nutrient in plant tissue, divided by the average concentration of the nutrient in water, gives a demand:supply ratio. The higher the ratio, the greater scarcity of the particular nutrient. Vallentyne (1972) found that regardless of season, P was the most scarce nutrient relative to plant needs (Table 4).

Thus, on the average, in fresh water, the measured ratios suggest that P is limiting, whereas in the sea such data indicate N to be most limiting. However, the nutrient-limiting growth rate may vary greatly from one freshwater subsystem to another and in time, depending mostly on the enrichment level. At the same time, P is probably the most important factor in the long-term sequence of lake stages. These finer points will be covered in the section on phytoplankton and eutrophication.

Table 4. *Ratio of plant-to-water concentrations for important plant nutrients in a variety of freshwater habitats around the world*

Nutrients	Winter	Summer
P	100 000	800 000
N	20-25 000	100-125 000
C	5-6 000	6-7 000

Source: After Vallentyne (1972) p. 107.

Shelford's "law" of tolerance is really the law of "too much" (Odum 1959). The occurrence of an organism can be controlled by qualitative or quantitative deficiency or excess in any one of several factors that may approach the limits of tolerance of that organism. The law relates, of course, to such environmental variables as toxic inhibitors, quality or quantity of food supply, inorganic nutrients, temperature, dissolved solids or salinity, light, water velocity, oxygen, and hydrogen-ion concentration (pH). The term organism here refers to a species population and not to an individual.

Adaptation

For each environmental factor an organism has a "built-in," or genetic, tolerance. For example, organisms have requirements for such elements as iron, copper, zinc, and manganese. These micronutrients, or trace metals, are required in small concentrations by microorganisms, which supply them, in turn, to higher trophic levels in the food web. However, when concentrations of these elements become too high they become toxic, at some concentration, to all forms of life, as illustrated in Figure 34.

An organism whose optimum for activity, growth, and survival is in a range of factor concentration denoted by A would be lost from the system and replaced by a species whose optimum is in range B if a permanent such change in that environmental factor occurred. If the optimum range of all pertinent environmental factors, including such biotic factors as competition and symbiosis, could be determined for a species, the *niche* for that species would be known. However, from the standpoint of water-quality control, the physicochemical factors are usually of concern and most easily defined.

Figure 34. Hypothetical response of two populations to a required element.

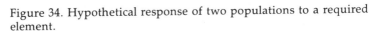
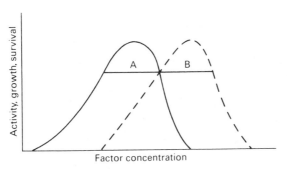

Each species is adapted through genetic selection to a set of environmental conditions. Further genetic adaptation to changed environmental conditions can only proceed slowly through several generations. However, within the limits of genetic adaptation, an organism can adapt physiologically. For example, fish can acclimate to temperature changes and algae can adapt to low and high light intensity. The organism that can best maintain enough growth to offset its respiratory demands in the face of changing environmental factors beyond its optimum range should be able to maintain the most stable population. That capacity would have survival value. Thus, a fish species can tolerate much higher temperatures in summer than in winter, having taken its cue from gradually changing environmental temperatures and light.

6

NUTRIENT CYCLES

Phosphorus

A discussion of the cycles of nutrients should logically begin with phosphorus (P), because it is usually the most limiting factor in fresh water, as mentioned earlier. First a general cycle will be discussed with a brief description of the important processes involved. Then the processes involved at the mud-water interface will be described in more detail. Finally, the principles involved will be applied to laboratory and open-lake conditions by reviewing experiments that have been performed.

The P cycle can be most simply represented as shown in Figure 35. The ultimate source of P to aquatic ecosystems is phosphate rock, whether through natural erosion or via waste inputs. P is utilized through plant and microbial uptake of DIP (dissolved inorganic phosphorus) through the processes of photosynthesis, chemosynthesis, and decomposition. Photosynthesis is the most important in the open-water areas; that is, the changes in DIP content are largely related to variations in plant uptake. The process involved in this uptake has been modeled and found to generally follow the Michaelis-Menten form as discussed previously. Subsequently, green plants are consumed by grazers and grazers, in turn, by predators. The excretion of DIP from plants and animals, as well as the death of plants and animals, leads to detrital particulate and dissolved organic P (POP, DOP), which is decomposed, in turn, by microorganisms releasing DIP. DIP is also released from POP and DOP via autolysis (Golterman 1972).

The remaining processes involve the sedimentation to and release from the bottom sediments - often referred to as the mud-water interchange. Sedimentation of detrital and living plankton results in a major loss of P from the open water, except possibly in rapidly flushed systems (rates greater than 0.1 per day). Much of the sedimented P is refractory and becomes part of the permanent sediments. Another fraction is mobile and under certain conditions can reenter the overlying water to furnish a feedback mechanism.

Where in the lake are these processes performed? Part of the cycle may be completed in the photic zone and/or within the metalimnion. However, the entire cycle includes the full extent of the water column and the littoral zone and is not complete until particulate matter has reached the hypolimnion and recirculated during turnover.

Although all these processes are important to the economy of P in a lake, the dynamics of P in a lake can be simplified for some purposes. For many systems the total P existing in the water, on the average, can be predicted rather accurately by knowing the inflow and outflow rates and the sedimentation rate. If sediments and macrophytes account for a considerable feedback of P, such a simplified approach may not work. Considerable progress has been made in understanding the P cycle of lakes, particularly in relation to the loading and trophic state of a lake as a result of P manipulation, which is of particular importance for the discussion here. This will be covered in detail later.

Figure 35. The aquatic phosphorus cycle with the transfers of greatest magnitude emphasized.

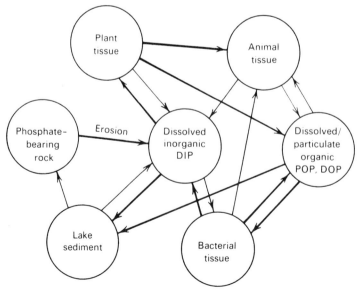

Reactions at the mud-water interface

Chemical oxidation-reductions at the mud-water interface, which includes the top few centimeters of sediment, are primarily controlled by the interface oxygen transport to the mud, which is a function of the hypolimnetic volume and turbulence (Mortimer 1942). During summer in a eutrophic lake, oxygen frequently declines to zero in the entire hypolimnion, but depletion occurs first at the mud-water interface because of the concentration of decomposition processes there. The mud surface becomes reducing and iron is reduced from Fe^{3+} to Fe^{2+} (ferric to ferrous). During turnover, oxygen is transported to the mud surface and iron is oxidized to the ferric state. In unstratified, shallow, eutrophic lakes, deoxygenation is possible at the mud surface under quiescent conditions. In oligotrophic lakes, surface sediments are not as reduced and remain highly oxidized because the hypolimnetic oxygen content is not depleted at any time. Mortimer (1941, 1942) has shown the contrasting state of sediments in oligotrophic Ennerdale and eutrophic Esthwaite Water in the English Lake District (Figure 36). Although the surface sediments are oxidized in Esthwaite water, they are more reduced throughout the sediment profile than in Ennerdale.

Figure 36. Vertical distribution of oxidized and reduced conditions in the sediments of two English lakes in winter. (Modified from Mortimer 1942 with permission of Blackwell Scientific Publishers Ltd.)

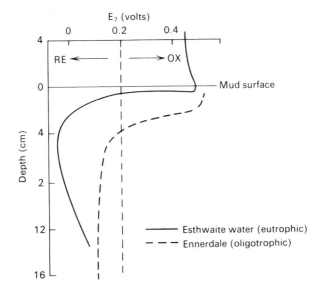

The release rate of PO_4 into the overlying water under such anaerobic conditions has been shown in Lake Mendato to be about 10 times greater than from the same sediments under aerobic conditions. The rate of release depends upon the area:volume ratio of the lake – the higher the ratio, the greater the release – and the gradient of sediment interstitial P with that of the overlying water. However, some release still occurs under aerobic conditions (Lee 1970), particularly if the pH is high. An anaerobic release rate in soft-water lakes can be around 3–5 mg P m^{-2} per day (McDonnell 1975), whereas rates as high as 50 mg P m^{-2} per day have been measured in hard-water lakes (Bengtsson 1975).

Although release rates can contribute a large amount of P to the overlying hypolimnetic water, considerable doubt exists as to the availability of that sediment-originated P in the economy of the lake. This is because the thermocline usually acts as a permanent barrier to vertical transport of materials (see Chapter 3).

Ferric iron and its hydroxy complexes strongly react with PO_4 and through precipitation and sedimentation remove P from the water column. $Fe(OH)_3$ is most efficient at removing PO_4 at a pH around 6 as shown in Figure 37 (Ohle 1953). The form of precipitated PO_4 may not be ferric phosphate, but rather a ferric hydroxy complex onto the

Figure 37. Relative removal of PO_4 from solution with $Fe(OH)_3$ versus pH. (Ohle 1953)

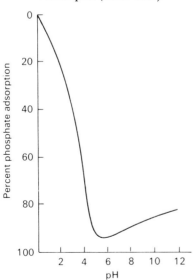

surface of which PO_4 is sorbed. Nevertheless, Fe^{3+} readily removes PO_4 (Lee 1970). Through the process of reoxygenation of the water column at turnover, Birch (1976) determined that in monomictic Lake Sammamish 90% of the P released during the stratified anaerobic period was quickly resedimented with oxidized iron. Although much P was released into the hypolimnion in late summer and early fall, little was left by the time phytoplankton grew in March or April.

PO_4 is also readily adsorbed onto organic particulate matter and absorbed by plankton, which can then settle out. Other metals are very capable of tying up PO_4, such as Mn^{2+}, Al^{3+}, and Ca^{2+}. The presence of the appropriate complex depends on the oxidation state of the metal and the solubility on pH (Stumm and Leckie 1971). Alum – $Al_2(SO_4)_3$ – is used as the principal complexing agent for removing PO_4 in sewage effluent (tertiary treatment or advanced waste treatment, AWT) and has been used to recover eutrophic lakes with some success.

Thus, reducing conditions in the top mud layer can release PO_4 (if complexed with Fe), but upon reoxygenation of the water column, the PO_4 is again tied up with oxidized metal (Fe) complexes and resettles. Whether this chemical release and resedimentation result ultimately as a net source for phytoplankton and a buffer to changes in the outside supply depends upon the extent of mixing between bottom waters and the photic zone. Sediment-released PO_4 can be a significant source if erosion of the thermocline downward during windstorms entrains and mixes released PO_4 upward in summer when light is available to stimulate phytoplankton uptake. This has been shown in Lake Ballinger when a storm in August resulted in disruption in the thermocline (Figure 38). Stauffer and Lee (1973) showed that a thermocline sinking of 1 m in Lake Mendota resulted in a doubling of the epilimnetic P content and contributed significantly to summer algal blooms. Analysis of continuous temperature data from Lake Ekoln, Sweden, showed slight thermocline migrations during the stratified period in 1970. An example is shown in Figure 39 in which the P content of the epilimnion increased (calculated) about 15%.

In a temperate, nonstratified (polymictic) lake, complete water-column circulation frequently occurs during summer, which is followed by stagnation, rapid oxygen depletion, and release of PO_4 before the mixing and reoxidation cycle recurs. P released from sediments in such a lake would continually be available to phytoplankton in the lighted zone. There are many shallow lakes in which such a mechanism could be significant but the process is not well defined.

To emphasize the importance of sediment as a P source, a recovery study on Lake Trummen in Sweden should be mentioned, a lake that probably operates in such a polymictic fashion. The lake is only 2 m deep, and core studies showed that although the sedimentation rate decreased over the 10 000-year history of the lake, it recently increased with sewage input from a rate of 0.2 to 8 mm per year. The sediment P content correspondingly increased during 1920–58 (from 50 to 800 mg l⁻¹) as did Zn and Cu content. Results from sediment nutrient release studies are shown in Table 5 (Björk et al. 1972).

The recently deposited sediments were obviously big sources for nutrients; even under aerobic conditions P release was significant as previously indicated. Interestingly enough, the release under aerobic conditions was about one-tenth that of anaerobic. To recover the lake, sewage was diverted, but the lake did not respond over a 10-yr period – the sediment apparently maintained the high rate of P supply for plant growth. Then over a 2-yr period 1 m of sediment (the nutrient-rich layer) was dredged from the lake and treated with alum, which produced a supernatant with 30 μg l⁻¹ P that was allowed to return to

Figure 38. Dissolved P increase in the epilimnion of Lake Ballinger, Washington, as a result of an August storm with a subsequent response in plankton chlorophyll.

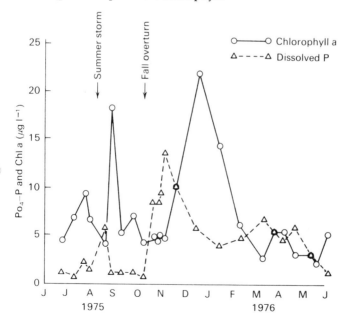

Table 5. *Release of N and P in containers with water and sediment under aerobic and anaerobic conditions (mg m⁻² day⁻¹)*

	Aerobic		Anaerobic	
	PO_4-P	NH_4-N	PO_4-P	NH_4-N
Black gytta deposited in 1960s	1.7	0.0	14.0	73.0
Brown gytta deposited 1000 yr BP, depth > 40 cm	0.0	0.0	1.5	0.0

Source: Björk et al. (1972).

Figure 39. Temperature profiles from Lake Ekoln, Sweden, preceding and following one of eight thermocline disturbances during the summer of 1970. Thermocline position estimated by extrapolation to approximate the overall thermocline displacement (6.5 m). Temperature was continuously monitored at 1 m intervals. (Data from National Swedish Environmental Protection Board)

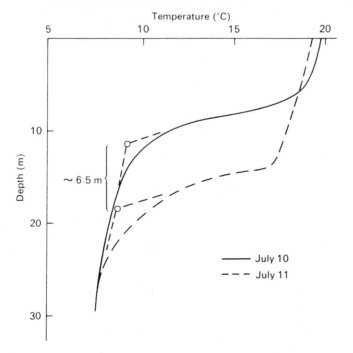

the lake. The lake has subsequently shown impressive signs of recovery, which will be discussed further in the section on control of eutrophication.

If the lake is deep, remains stable, and hypolimnetic oxygen does not deplete during the summer, the sediments operate largely as a sink. However, if the lake is shallow enough not to stratify and yet does not permanently deoxygenate, the sediments may still act as a source even though release from the usually aerobic sediments is slow and the sediment still accumulates more P than is released. A potential mechanism for transferring P from sediments in littoral areas is through death and decomposition and possibly excretion by rooted macrophytes (Schultz and Malueg 1976; Carpenter and Adams 1978), which will be considered further in the following section. It should be noted that PO_4 may also be released from aerobic littoral sediments under photosynthetically caused high pH if iron is the controlling element.

Combined chemical and biological processes that establish P equilibrium

Laboratory and field experiments have demonstrated that lakes tend to reestablish equilibrium levels of P in the water in a relatively short time following manipulation of the inputs – often faster than water exchange rates would indicate. In laboratory microcosm experiments, responses to added radiophosphorus (^{32}P) were observed (Figure 40). This experiment was one of the first to emphasize the rate of P cycling in ecosystems and the potential ability of the system to reestablish previous equilibrium levels (Hays and Phillips 1958).

Figure 40. Distribution of ^{32}P among three forms after addition to microcosms. DIP, dissolved inorganic P; DOP, dissolved organic P; PP, particulate P. (Hays and Phillips 1958)

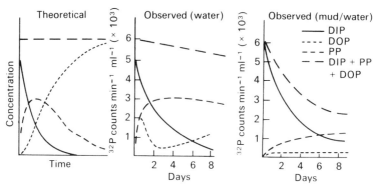

The principal points that this experiment shows, as indicated by Hays and Phillips (1958), are as follows:

1. The reaction did not go to completion as predicted; all DIP was not converted to DOP (dissolved organic phosphorus).
2. Bacteria held much of the P as PP, (particulate phosphorus), which continued to be exchangeable with the water phase.
3. The loss of P was much greater with mud present, which competed with bacteria for DOP.
4. An equilibrium process was established between water and solid phases as follows:

   DIP $\rightleftharpoons$ PP $\rightleftharpoons$ DOP

5. The equilibrium level of ^{32}P was about 0.1 of the initial level, following an exponential decline as follows:

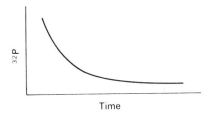

These laboratory observations were repeated in a small Nova Scotia lake to which 11 mc of ^{32}P was added as $KH_2{}^{32}PO_4$. The level decreased by 80% after 2 wk. The concentration peaked in plankton after 4–16 hr and in fish after 3–4 days (Hays and Phillips 1958).

Experiments in Crecy Lake, New Brunswick (20 ha, mean depth 2.4 m), with inorganic fertilizer showed similar results. The lake was fertilized three times, which theoretically raised P to 390 μg l^{-1}, N to 210 μg l^{-1}, and K to 270 μg l^{-1}. The results of fertilization in 1951 and 1959 and successive years are shown in Figure 41 (Smith 1969).

From such experiments, the turnover times (time required for loss from a phase), or the turnover rate (1/turnover time), can be estimated, as was done for Crecy Lake (Table 6).

These experiments illustrate the dynamic character of the P cycle. P is absorbed very rapidly from the water because it is in great demand by microorganisms and is effectively sorbed by other particulate matter and chemical complexes. The loss from microorganisms is slower than from water but faster than from sediment. At the same time there is continual replenishment of DIP to the water phase from these solid phases such that the amount of P that can actually be measured at any one time may be very small (usually less than 10 μg l^{-1} in lakes in summer), but the rate of transfer to and from the water phase is very great. This rate of replenishment then is the supply of P available to the producer organisms. Concentrations of P at a point in time are consequently not an adequate indication of supply for algal production.

Figure 41. Effect of fertilizing Crecy Lake with P, N, and K. Concentrations of dissolved inorganic and organic phosphorus (DIP, OP) during years of fertilizing and succeeding years are shown. (After Smith 1969)

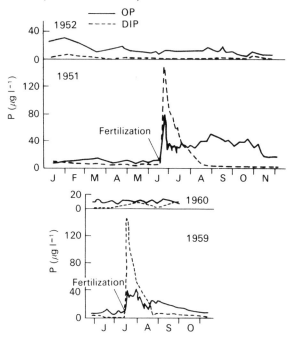

Table 6. *Turnover time of P in water and solids of Crecy Lake estimated from observed loss rates*

Fertilization	Percent loss rate day^{-1}	Turnover time in days	
		Water	Solids
1st	5.9	17	176
2nd	5.6	18	248
3rd	5.3	19	394

Note: Turnover time (days) = (amount in phase)/[loss rate (per day)]
Source: Smith (1969).

Fate of added P to lakes

What is the fate of P added to lakes from external sources and the role of the littoral region? Although the littoral contribution has been recently stressed, Hutchinson (1957) conducted an early study on Linsley Pond, Connecticut, to which ^{32}P was added June 21 and results observed August 1–15 by measurements of the tracer in different areas. This experiment, which involved the whole lake, illustrated the interaction between processes in the epilimnion, the littoral region, and the hypolimnion on the transfer of P. The following is an accounting of the transfer rates of P in kilograms per week among general compartments (Hutchinson 1957, p. 748):

Epilimnetic increase was	0.26
Loss to hypolimnion was	1.55
Transport from littoral to epilimnion is the sum	1.81
Hypolimnion considered a sink and showed gain of	3.75
Gain from epilimnion was	1.55
Gain from sediment was	2.20

The assumption here was that the hypolimnion acted as a sink. The gains in the epilimnion may have been partly from the hypolimnion through the thermocline disruption processes mentioned earlier. However, the author stressed the transport from the littoral region, a source that has been emphasized rather strongly in recent studies. Cooke et al. (1978) have shown that much of the summer supply of P to the epilimnion of twin Lakes, Ohio, originated from the littoral. In their study epilmnetic P could not have originated from hypolimnetic muds because that bottom was sealed with alum flock.

Modeling P exchanges

To develop models to predict the response of aquatic ecosystems to manipulation of the P input, the pertinent pools and transfer rates among the pools must be determined. Figure 42 shows a reasonable approximation of pool sizes and transfer rates in the three portions of a stratified lake (Stumm and Leckie 1971).

Note that the transfer rates are expressed as μg l^{-1} per day and the pool sizes as μg l^{-1}. At steady state, about 30% of the uptake demand by the phytoplankton (P supply rate) comes from bacterial decomposition in the epilimnion, whereas 20% comes from the hypolimnion and the remainder from inflow. The hypolimnetic source implies thermocline erosion following a storm or breakup at overturn. Likewise, 30% of the P loss rate continuously rains into the hypolimnion via sinking consumer fecal pellets and dead cells. The principal omissions in this diagram are regeneration from dead cells via autolysis, which is not a function of bacteria, and the excretion of DIP by grazing zooplankton, probably the most important regenerative process, is also omitted.

In a large lake like Sammamish (20 km²) with a 2-yr water residence time, the inflow rate tends to furnish a minor share of the required P supply demanded for average rates of phytoplankton photosynthesis. More than 80% of the P supply during the stratified period was estimated to have come from regenerative processes, of which zooplankton is probably the major contributor (Welch and Spyridakis 1974).

Figure 42. Hypothetical steady-state model of phosphorus cycling in a lake. (After Stumm and Leckie 1971)

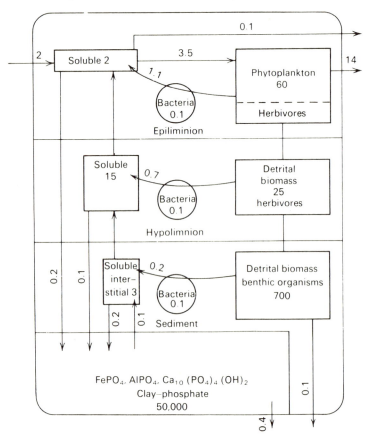

A further point in Figure 42 is that chemical regenerative mechanisms in sediment account for only 5% of the biological contribution (0.1 vs. 2.0 μg l^{-1} per day). In a shallower and possibly anaerobic lake, this sediment contribution is bound to be more significant.

Sediment as sink or source

The problem of whether sediments act as a source to maintain P in overlying water or as a sink includes several arguments. The principal pro-sink arguments are as follows:

1. Aerobic sediment has been shown to completely absorb PO_4, and algae did not outcompete sediment for PO_4 in short-term emperiments (Fitzgerald 1970).
2. There is usually a net sedimentation rate of PO_4. Lake inflow PO_4 > outflow PO_4. The retained fraction can vary widely but is usually greater than 0.5 (Vollenweider 1975). This can also be illustrated by comparing the water and P residence times. These are 2 and 0.77 yr, respectively, in Lake Sammamish.
3. P is released in anaerobic conditions but is reprecipitated with vertical mixing and oxygenation and is usually not available to the trophogenic zone.

The principal pro-source arguments are as follows:

1. Even aerobic sediment yields P slowly if pH is high and macrophytes can slough decomposable matter and excrete P (Wetzel 1975); consequently, the littoral zone can be a source of P in shallow lakes or lakes with relatively large littoral areas.
2. A high percent ($\sim$ 80%) of fixed P is regenerated in the photic zone prior to sedimentation (Golterman 1972; Birch 1976). Carp are efficient excretors of PO_4 (Shapiro et al. 1975). However, regeneration cannot sustain a constant supply indefinitely if the input has decreased.
3. Theoretically, a large release of P, which can be available for growth, occurs from sediment in shallow, frequently destratified (polymictic) lakes during summer. P content in interstitial water is often 50 or more times that in overlying water. Thus, frequent destratification avails P to the productive zone. The depth of the active releasing zone is 5–10 cm. The high ratio of area of sediment surface:water volume results in a high P content in the water (Stumm and Leckie 1971).
4. Thermocline sinking has been demonstrated in Lake Mendota, a 20 m deep lake, following storms, which doubled the epilimnetic P content (Stauffer and Lee 1973).

Obviously, the role of sediments in the P budget of a lake is very complex, and, therefore, none of these points are answers in themselves as to whether the sediments are going to dominate lake-water P levels. Some of the processes suggested are more important in time and space than others. To determine the importance of various sources (internal and external) is difficult enough in a static situation. However, whether that source rate remains stable in the face of manipulating one or two external (or internal) sources is even more important and very poorly understood at present. Although it is true that certain generalities can be made regarding sediment release under aerobic and anaerobic conditions, such factors as the relative amount of regeneration going on in the pelagic zone, the retention capacity of lakes, and the littoral input, to name a few, contribute considerable lake-to-lake variation. When predicting changes in the level of P in a particular lake any more accurately than ± 50%, it must be realized that lakes process P very much on an individual basis and much must be known about the individual lake in question to predict, with any greater accuracy, any changes that can be expected from manipulation.

Nitrogen cycle

The general cycle of N in an aquatic ecosystem is shown in Figure 43. The cycle is very complex, and many of the transfer processes and associated pool sizes are important not only to aquatic productivity

Figure 43. The aquatic nitrogen cycle.

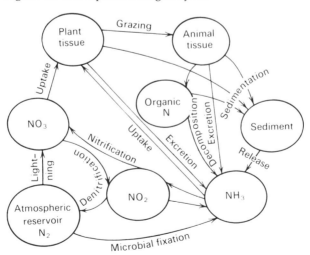

but also to environmental quality in general and human health. Two large biological source and sink processes occur with N that do not occur with P, which greatly influences the significance of N in the control of productivity and waste treatment. These processes involve the transfer of N from atmospheric N_2 through microbial fixation and its return to the atmosphere via NO_2 and denitrification. The processes of nitrification (oxidation of ammonia) and denitrification are reactions that would proceed without biological mediation but at a slow rate. Microorganisms greatly speed the rate of reaction and at the same time capture the energy available in the reduced compounds through an ordered series of cellular enzyme catalyzed reactions. Because the energy sources are inorganic, the organisms are referred to as chemolithotrophs (Brock 1970).

Pool concentrations

Nitrogen is most abundant as N_2, comprising nearly 80% of the atmospheric constituents. However, before this form is available to plants it must be fixed either biologically or through a lighting energized reaction. NO_3 is the most common form that can be used by aquatic plants, and its content ranges from a trace when productivity is high and all available N has been absorbed into plant tissue to concentrations usually between 500 and 1000 μg l^{-1} during periods of no utilization. Concentrations well over 1 mg l^{-1} can occur but are usually associated with waste inputs. NH_3 becomes abundant in the absence of oxygen or in very enriched waters, but it is usually less abundant than NO_3. Because NH_3 is a more reduced form, it is often preferred over NO_3 by plants. NO_2 is not used and in fact can be rather toxic. Because it readily oxidizes to NO_3, it is usually not present in any appreciable amounts.

Nitrification

Nitrification is the process by which NH_3 is transformed into NO_2 and finally into NO_3. The process occurs only under aerobic conditions. Organisms that commonly perform the transformations are *Nitrosomonas* and *Nitrobacter*. Although the processes are energy yielding, as shown below, the yield is rather low compared to other transformations in the cycle (Delwiche 1970). The reactions of nitrification are:

$$2 \ NH_3 + 3 \ O_2 \rightarrow 2 \ HNO_2 + 2 \ H_2O + energy$$
$$2 \ HNO_2 + O_2 \rightarrow 2 \ HNO_3 + energy$$

Denitrification

This process occurs only in the absence, or near absence, of oxygen. A common denitrifying organism is *Thiobacillus denitrificans* and the associated reaction is:

$$5\ S + 6\ NO_3 + 2\ H_2O \rightarrow 5\ SO_4 + 3\ N_2 + 4\ H + energy$$

That organism is, of course, a chemolithotroph.

Heterotrophic bacteria, such as *Micrococcus, Serratia, Pseudomonas,* and *Achromobacter*, are also denitrifiers when oxygen concentrations are low. As they are facultative anaerobes, they can also carry on normal aerobic respiration and are part of the normal flora in sewage (Christensen and Harremoes 1972).

In denitrification, the reverse process of nitrification, the bacteria reduce NO_3 to NO_2 and in turn form molecular gaseous N_2. This, then, represents the mechanism of N loss from ecosystems and can be used as a treatment method for the removal of N from waste water. However, the environment must be aerobic for nitrification and anaerobic for denitrification. The necessary alteration of aero/anaero-biosis in order for denitrification to represent a significant N loss has implications in the N balance and nutrient limitation in lakes as well as in waste-water treatment. The more productive a lake, the better chance for oxygen to reach low levels, which would allow denitrification. For waste-water treatment, an aerobic period must precede the anaerobic period in order for NH_3 to be converted to NO_3, which can then be denitrified without oxygen. Denitrification to NH_3 from NO_3 also occurs, but little seems to be known about that process.

Fixation

Nitrogen fixation is an energy-consuming aerobic process carried on in aquatic environments by such bacteria as *Azotobacter* and *Clostridium* and by the blue-green algae *Nostoc, Anabaena, Anabaenopsis, Gleotrichia,* and *Aphanizomenon*. These organisms usually dominate in enriched lakes in summer. Nitrogen fixation can represent a significant input of N to an ecosystem. Measured rates for *Anabaena* range from 0.04 to 72 $\mu g\ l^{-1}$ per day. Such rates could represent a severalfold turnover in dissolved inorganic N in enriched systems, because NO_3 is usually depleted to very low levels or is undetectable when blue-green fixers appear. Horne and Goldman (1972) have shown that blue-green fixation contributed 43% of the annual N input to Clear Lake, California.

N fixation is performed by the heterocystis blue-green algae. The abundance of these extralarge cells in the filaments or chains of the above N-fixing blue-greens usually increases with NO_3 depletion (Horne and Goldman 1972). Because fixation is an energy-demanding process, it becomes advantageous only when NO_3 and NH_3 are no longer available.

This source of N for the N fixers also becomes available for other algal species as the cells of the N fixers decompose. In many eutrophic lakes *Aphanizomenon* and *Microcystis* usually do not occur together in large masses but rather alternate. N availability may be the cause for this alternation. Also, the availability of N can be a cause for the succession to blue-green algae, particularly the heterocystis blue-greens.

Just as denitrification and N loss can increase in significance as productivity increases (because of oxygen depletion), fixation also increases with productivity because of the greater depletion of NO_3. Fixation rates of the planktonic blue-greens in enriched Lake Erie have been determined to range from 4.2 to 230 n moles acetylene per mg N per hour and from 0.69 to 25 in less rich Lake Michigan (Howard et al. 1970). Horne and Goldman (1972) have measured a maximum rate of several hundred n moles per liter per hour in highly enriched Clear Lake.

Few adequate balances of N are available from aquatic systems. Brezonik and Lee (1968) have developed a rather complete balance of N income and loss in Lake Mendota (Table 7). In that case the gain from fixation was very similar to that lost by denitrification, although neither was the most important source or sink. Sedimentation was estimated by differences between loss and income. Another source and sink,

Table 7. *An N balance from Lake Mendota, Wisconsin*

Income		Loss	
Source	Percent	Sink	Percent
Waste water	8.1	Outflow	16.4
Surface water	14.7	Denitrification	11.1
Precipitation	17.5	Fish catch	4.5
Groundwater	45.0	Weeds	1.3
Fixation	14.4	Sedimentation	66.7
Total	100	Total	100

Source: Modified from Brezonik and Lee (1968).

respectively, were marsh drainage and groundwater recharge for which they had no values.

Implications to nutrient limitation

The availability of N and P to the primary producers in freshwater ecosystems is markedly different because of unique processes in the N cycle, together with the differing chemical behavior of N and P. The reiteration of some of these points should give better insight as to which nutrient is apt to be most limiting.

1. In a freshwater lake the residence time for an incoming quantity of N is longer than that for P, because the strong sorptive capacity of inorganic metal complexes and organic particulate matter for PO_4 tends to remove P to the sediments. Although some sedimented P can become available again in anaerobic environments or through rooted macrophyte uptake and loss, in most cases the overall efficiency of such recycling does not seem to be very high. N, on the other hand, in the form of NO_3 and NH_3, is much more soluble and is not readily sorbed by the inorganic complexes as is PO_4. Besides, NH_3 can also be released from anaerobic sediments.

2. Although N can be lost from aquatic systems via the biological process of denitrification, which is unique to the N cycle, this supposedly can happen only in anaerobic and thus highly enriched waters. Although this process could contribute to N limitation in such highly enriched environments, it should not greatly contribute to an N shortage in waters of low or moderate enrichment.

3. The atmosphere provides a ready source of N to N-depleted systems through the biological fixation of N_2. This occurs in aerobic environments and needs only the presence of the right kind of microorganisms. When available N becomes depleted, these organisms can dominate and thus provide a very large part of the N supply to some systems, which are usually very productive. In low or moderately productive systems N availability has not declined enough to offer an advantage to N fixation. Because it is an energy-consuming process, its only advantage is a scarcity of NO_3 and NH_3. Another fact that restricts the supply of N through fixation is a rather low maximum rate of N cell replacement, 0.05 per day (Horne and Goldman 1972).

4. Yet another original source of N that does not occur in the P cycle is rain. Although rainwater contains P and has been shown to be important in lakes in phosphorus-poor watersheds (Schindler 1974a), the NO_3 in rain is a rather constant phenomenon, having been transformed from the N_2 reservoir in the atmosphere by lightning. In a temperate area the N in rain is about 6 kg ha^{-1} per year for a 75 cm level of precipitation (Hutchinson 1957).

Overall, it is obvious that fewer sources exist for P than for N and that the sediment is probably a more efficient remover of P than N in aquatic ecosystems. Moreover, from the discussion of the processes in the N cycle, N is most likely to become limiting in highly enriched systems. Therefore, the greatest control on productivity would most likely result from control of P inputs with possibly some added control from N restriction in systems already highly enriched.

This reasoning does not seem to be supported in marine systems where N is usually the most limiting nutrient (Ryther and Dunstan 1971). An explanation for this is not entirely clear, although N fixation is apparently rather low in the sea and the recycling of N in the water column is much slower than that for P. The recycling process in the sea is the reverse of the process in freshwater lakes (Schindler 1974a; Birch 1976), where particulate matter becomes relatively richer in P than N with depth in the water column. It could be that the mixed layer in the ocean, which is deeper, allows a more complete remineralization of settling particulate matter and thereby consistently low dissolved N:P ratios, whereas the usually shallower depth of lakes results in the trapping of P in sediment before it can all be released, and, consequently, much higher N:P ratios predominate. In both fresh water and seawater, waste-water inputs are relatively more enriched with P than N compared to the receiving water.

Sulfur cycle

The sulfur cycle is interesting from the standpoint that several significant water-quality changes result from man's waste and involve processes in the S cycle. However, S itself is almost never a limiting factor in aquatic ecosystems. The normal levels of SO_4 are more than adequate to meet plant needs. The cycle is shown in Figure 44 with some of the bacteria and associated processes that are responsible for the indicated (numbered) transformations in the cycle.

Odorous conditions are easily created when waters are overloaded with organic waste to the point that O_2 is removed. Then SO_4 is the electron acceptor often used for the breakdown of organic matter and step 1 in the cycle produces H_2S and a rotten-egg smell. If NO_3 is available, N-reducing bacteria will dominate. The production of H_2S, which is toxic, has been linked with fish kills.

Thiobacillus ferroxidans contributes to acid mine water by oxidizing FeS (ferrous sulfide), resulting in H_2SO_4 and pH near 1.0. Such bacterial activity may cause 80% of such acidity. The same process occurs with *T. thiooxidans* and is responsible for pipe corrosion (Brock 1970).

Of considerable interest is the occurrence of "plates" of sulfur bacteria in lakes (Brock 1970), either photosynthetic, facultative anaerobic purple or green sulfur bacteria or aerobic, colorless sulfur bacteria. Regardless of type, they are rather restricted to intermediate layers, often the metalimnion. Photosynthetic bacteria need light as well as H_2S. Although H_2S may be abundant at depth, light is not, and the abundant light at the surface is of no help because H_2S is unstable in the presence of O_2. So these organisms occur at a rather restricted depth interval (plate) where both light and H_2S are adequate. Obviously in aerobic lakes they do not occur.

The aerobic, nonphotosynthetic bacteria have much the same problem of restriction. Because they require both O_2 and H_2S, their ap-

Figure 44. The sulfur cycle (modified from Brock 1970) and associated bacteria-mediated reactions. (Klein 1962)

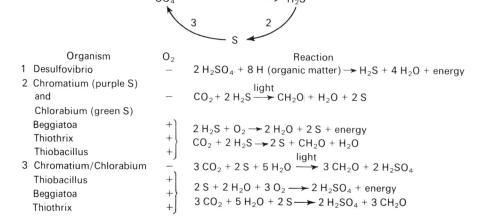

	Organism	O_2	Reaction
1	Desulfovibrio	−	$2 H_2SO_4 + 8 H$ (organic matter) $\rightarrow H_2S + 4 H_2O$ + energy
2	Chromatium (purple S) and Chlorabium (green S)	−	$CO_2 + 2 H_2S \xrightarrow{\text{light}} CH_2O + H_2O + 2 S$
	Beggiatoa	+	$2 H_2S + O_2 \rightarrow 2 H_2O + 2 S$ + energy
	Thiothrix	+	$CO_2 + 2 H_2S \rightarrow 2 S + CH_2O + H_2O$
	Thiobacillus	+	
3	Chromatium/Chlorabium	−	$3 CO_2 + 2 S + 5 H_2O \xrightarrow{\text{light}} 3 CH_2O + 2 H_2SO_4$
	Thiobacillus	+	$2 S + 2 H_2O + 3 O_2 \rightarrow 2 H_2SO_4$ + energy
	Beggiatoa	+	$3 CO_2 + 5 H_2O + 2 S \rightarrow 2 H_2SO_4 + 3 CH_2O$
	Thiothrix	+	

pearance is restricted to the interface of declining O_2 and H_2S levels, often at the bottom of the metalimnion.

Carbon cycle

The carbon cycle is the one most frequently studied in ecosystem trophic transfer work. It too is strongly implicated in water-quality change and in the control of productivity and alkalinity. Some important pools and pathways in the cycle are shown in Figure 45.

The availability and role of inorganic carbon in aquatic productivity is a very important and controversial subject. To understand some of the critical issues regarding productivity limitation and algal species succession, certain aspects of CO_2 will be discussed.

CO_2 system

The supply of DIC (dissolved inorganic carbon) in water is of great importance to primary productivity because about one-half the dry weight of plankton algae is carbon. The CO_2 system is the principal source of DIC. Also, the availability of free CO_2, versus HCO_3^- or CO_3^{2-}, has been linked rather convincingly to the succession phenomenon in plankton algae (King 1970, 1972; Shapiro 1973; Shapiro et al. 1975). For these reasons some pertinent points of the CO_2 system will be reviewed.

$CO_{2(g)}$ constitutes about 0.03% of the atmospheric constituents, and atmospheric $CO_{2(g)}$ is readily soluble in water. CO_2 dissolves in water to produce carbonic acid:

$$CO_{2(aq)} + H_2O \rightleftharpoons H_2CO_3$$

H_2CO_3 is dissociated according to the following reactions:

$$H_2CO_3 \rightleftharpoons H^+ + HCO_3^-$$
$$HCO_3^- \rightleftharpoons H^+ + CO_3^{2-}$$

As algae consume CO_2 from the system, more $CO_{2(aq)}$ can be made available from the following two reactions:

$$H_2CO_3 \rightleftharpoons CO_2 + H_2O$$

which is the inverse of the first hydration reaction and occurs at pH < 8. The second:

$$HCO_3^- \rightleftharpoons CO_2 + OH^-$$

occurs at pH > 10. As indicated in the last equation, as CO_2 is removed from HCO_3^- the OH^- is released, causing a rise in pH. This is an important phenomenon for continued photosynthesis because as the pH increases, the concentration of free $CO_{2(aq)}$ decreases.

All natural waters contain free CO_2, HCO_3^-, and CO_3^{2-} in equilibrium with each other and close to equilibrium with atmospheric CO_2. The relative proportion of each of the inorganic forms in water is a function of pH. This is shown in Figure 46 for a system with constant total carbon or C_T ($CO_2 + HCO_3^- + CO_3^{2-}$). Photosynthesis and res-

Figure 45. The carbon cycle with some important biomediated processes.
1   This process is normal aerobic respiration: $CH_2O + O_2 \rightarrow CO_2 + H_2O$.
2   The same respiratory process is performed by microorganisms through decomposition.
3   Bacterial decomposition is also performed by anaerobic respiration. Since anaerobic metabolism is incomplete, only a portion of the carbon goes to CO_2. The remainder is contributed to the dissolved organic pool (DOC), which is slowly metabolized to CO_2. Some anaerobic processes in which a carbon by-product other than CO_2 is produced are performed by methane bacteria, which can carry on the following overall reaction (Brock 1970):

$$C_6H_{12}O_6 + X\,H_2O \rightarrow Y\,CO_2 + 2\,CH_4$$

4   The principal source of free CO_2 is dissociated from the dissolved inorganic carbon (DIC) pool. This supply depends on the total inorganic carbon content, temperature, and pH.
5   Organism excretion contributes to the DOC pool
6   CO_2 is assimilated through photosynthesis into particulate organic carbon (POC) to initiate the cycle.

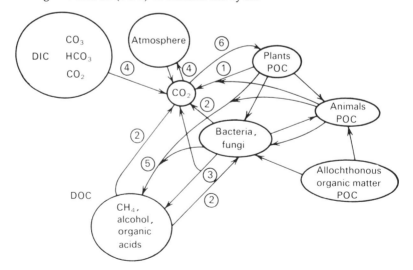

piration are two major factors that cause a significant departure from equilibrium of the system with the atmosphere. As the pH increases in productive environments toward midday and the concentration of $CO_{2(aq)}$ decreases to very low levels, the photosynthetic rate also decreases proportionately (King 1972).

The system provides a buffer against the addition of H^+ or OH^- through the above reactions as long as the supply of CO_2, HCO_3^-, and CO_3^{2-} remains the same. As a result, the pH remains constant. However, the effect of photosynthesis is to remove C_T and thus the buffering capacity of the system. If $CO_{2(g)}$ was absorbed from the atmosphere as fast as algal uptake, the pH would not rise, but the mere fact that C_T can be observed to decrease and pH increase is evidence that algal uptake does exceed the atmospheric resupply (Schindler 1971b; King 1972). Therefore, photosynthesis tends to be self-limiting with respect to the carbon system not only by reducing C_T but also by raising pH, thus reducing the concentration of free $CO_{2(aq)}$.

The rate of replenishment of $CO_{2(aq)}$ from H_2CO_3 and HCO_3^- is several times faster than necessary to match any conceivable algal demand rate (Goldman et al. 1971). Thus, the decrease in photosynthesis following continued CO_2 depletion and pH increase is not because the reactions are not rapid enough, but rather because the reduced free $CO_{2(aq)}$ concentration becomes increasingly rate limiting to photosynthesis. Although some algal species can utilize HCO_3^- directly (Goldman et al. 1971), such a shift in species as CO_2 becomes depleted could also tend to decrease productivity in the short term.

Figure 46. The distribution of the three forms of total C, when C_T is constant, as a function of pH. (modified from Golterman 1972)

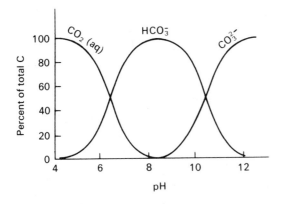

Study questions

1. Two lakes of different depths ($\bar{z} = 5$ and 20 m) thermally stratify and have anaerobic hypolimnia. Given that the external nutrient inputs are similar, *which lake* would probably show the greatest yield of phosphorus from sediments that would be available for use by plankton algae. Give *two reasons* for your choice.

2. Concentrations of inorganic soluble C, N, and P exist in the water in a weight ratio of $200:16:1$ ($518:35:1$ by atoms).
   a. Which element would be most apt to limit yield (i.e., maximum biomass)?
   b. Would growth rate (μ) be likewise limited by that nutrient if that ratio existed?
   c. *Explain*.

3. An internal process that can significantly increase the concentration of phosphorus in a lake water column is
   a. a rapid recycle rate through zooplankton grazing in the epilimnion
   b. destratification of a lake in autumn that was anaerobic during summer
   c. the anaerobic reduction of ferric hydroxy complexes in the sediment and subsequent diffusion
   d. the decomposition of settling particulate matter by bacteria

4. The turnover rate of phosphorus is
   a. 1/loss rate
   b. the same as the turnover time
   c. 1/time necessary to replace the quantity in a phase
   d. the time necessary to replace the quantity in a phase

5. Which of the following *does not* contribute strongly to phosphorus being the most limiting nutrient in fresh water?
   a. rapid uptake of PO_4^{-3} by organisms and subsequent sedimentation
   b. strong tendency for adsorption of PO_4^{-3} onto metal complexes
   c. a C:N:P ratio in plankton of $106:16:1$
   d. an atmospheric reservoir of N_2 and CO_2

6. Denitrifying and sulfate-reducing bacteria do *not*
   a. grow in anaerobic environments
   b. require the reduced forms of N and S

c. produce gaseous forms of N and S

d. require a carbon source

7. A lake has an average inorganic N:P ratio of 2:1 during the summer. All of the following could be observed in this lake except

a. a large input of N fixed from atmospheric N_2

b. a dominance of the plankton by blue-green algae

c. a consistently low pH

d. very high oxygen concentrations

8. Besides green plant photosynthesis, autochthonous production in aquatic environments can also result from the following energy and carbon sources and O_2 conditions

a. light and H_2S with CO_2 and without O_2

b. NH_3 with CO_2 and with O_2

c. S with CO_2 and with O_2

d. SO_4 with CO_2 and without O_2

e. H_2S with CO_2 and with O_2

f. all of the above

g. three of the above

h. four of the above

PART TWO

THE EFFECT OF WASTE ON POPULATIONS

7

PHYTOPLANKTON AND CONTROLLING FACTORS

The two major aspects of phytoplankton that should be considered are the productivity and biomass and the species composition. How much is being produced and what is it? Productivity and biomass can be measured by chemical or physical procedures (Vollenweider 1969b) and related to environmental factors without knowledge of the major organisms responsible. Although productivity and/or biomass may be all that one needs to know in certain situations, as will be seen later in this section, the kinds of organisms that are present is the major part of the problem in many cases. The trophic state of a lake can be defined quantitatively, based on measurements of productivity and biomass. Thus far, however, quantitative criteria for species composition have not been defined for the trophic state of lakes.

Although it is interesting and useful to know how and why which phytoplankton change seasonally in a particular lake, it is much more important to water-quality management to know how the average state of lakes in general changes with an increasing or decreasing load of nutrient, acidity, or sediment. The response of lake plankton to external manipulation is one of the most interesting subjects in ecology. Although small-container experiments (bottles, bags, etc.) give valuable information about such responses, they are largely inadequate because of the many factors operating in a large lake that interact to buffer or magnify the response. Therefore, deliberate or accidental manipulations of whole lake systems (Edmondson 1972; Björk 1974; Schindler 1974a) have presented more reliable information in that regard. Nevertheless, an understanding of the relative importance of the principal environmental factors is necessary before large-scale manipulations can be fully appreciated.

General seasonal pattern

The general pattern shown for the sea (Figure 47) is quite similar to that for a large monomictic or dimictic lake. The nutrient content is high following fall overturn and may remain high during winter in a monomictic lake; in a dimictic lake it may increase again during spring overturn, providing a large ready supply for plankton algae. The large nutrient availability exists principally because little or no growth is occurring at low light and temperature. A spring diatom "outburst" occurs when light intensity reaches a level where gross photosynthesis exceeds respiration. Species with low temperature requirements are usually responsible for the outburst.

A midsummer minimum occurs in algal biomass and productivity largely because nutrients have reached a low and production-limiting level. The controlling effect of nutrients is particularly evident because light and temperature are at their maximums, thus the production potential is high. Grazing by herbivorous plankton is also occurring and may also represent a significant loss rate.

In autumn the significance of nutrients is again readily apparent because increased mixing, which results from lowered surface temperatures, brings up regenerated nutrients into the trophogenic (productive) zone. More nutrients are often available at that time because of the continual sinking of the metalimnion during summer. A fall outburst of diatoms often results from this in the ocean, as is also the case for poorly enriched freshwater lakes. In highly enriched lakes, how-

Figure 47. General seasonal cycle in oceanic phytoplankton and ecological factors. (Modified from Raymont 1963)

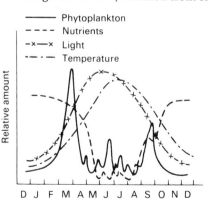

——— Phytoplankton
– – – – Nutrients
–x——x Light
–— ·— Temperature

Relative amount

D J F M A M J J A S O N D

ever, high concentrations of blue-green algae tend to develop in late summer and autumn or remain high during the entire summer.

The illustrated pattern (Figure 47) can be considered typical, but as environmental factors vary from lake to lake, this pattern takes different forms. For example, the pattern in Lake Sammamish (Figure 48) may range from a very large diatom outburst in April to a smaller maximum of a diatom – a blue-green mixture – in June and July. In both instances there is a small increase in October as a result of an increase in PO_4. During the last year of maximum sewage effluent input, the biomass in Lake Washington was greater and more sustained during summer. Since 1970 the pattern and biomass magnitude in Lake Washington (Edmondson 1972) has appeared more like that in Lake Sammamish.

Light effects on phytoplankton

Phytoplankton growth is affected by both the quality and quantity of light. The wavelengths of light are separated and the total amount diminishes exponentially as light passes through water. Therefore, it is not surprising that the effect of light on productivity and species occurrence has its most profound effects in aquatic systems. In very clear water productivity can be greatly inhibited in the top several meters by high intensity as well as by penetration of toxic ultraviolet rays and maximum productivity could occur at depths of from 10 to 15 m (Hendrey and Welch 1974). In highly productive systems the maximum is at the surface and the photic zone may be only a meter or two. In such a situation the productivity is limited by insufficient light.

Quality

Figure 49 shows the differential penetration of the various wavelengths of visible light through water. Infrared and ultraviolet light is readily absorbed by water and thus penetration is usually slight. Blue light is absorbed least and thus penetrates deepest. Lakes poor in nutrients are usually also poor in organic particulate matter, and as a result, they are very blue because blue light is allowed to penetrate to great depths. The poorer a lake is in nutrients, the bluer it usually appears.

The energy for photosynthesis is trapped largely by chlorophyll. Chlorophyll *a* has its greatest absorbance in the long and short wavelengths within the visible spectrum, specifically, the violet and red areas (Figure 50).

In spite of chl *a* having its greatest absorbance in the red and violet areas, the energy from blue and green light is greatest at a depth of 5 m in lake water. This suggests that plankton algae are poorly adapted to photosynthesize in water because most of the penetrating light is unavailable. However, algae contain many accessory pigments besides chlorophyll. These other pigments are useful to the plant in trapping energy in other wavelengths and passing this energy on to chlorophyll as follows (Schiff 1964):

carotenoids → phycocyanins → phycoerythrins → chlorophyll

Quantity

As stated above, light energy in the range 400–700 nm, or the visible part of the spectrum, is effective in photosynthesis. In spite of the specific absorption pattern of chlorophyll (Figure 48), photosynthesis varies with visible light intensity according to the relationship shown in Figure 51. At the low end of the intensity range photosynthesis is proportional to intensity, but at higher intensities photosynthesis levels off, or is "saturated" with light. At even higher intensities,

Figure 48. General pattern of phytoplankton biomass in Lakes Sammamish and Washington during the indicated years. (Data from Edmondson 1972; Welch 1977)

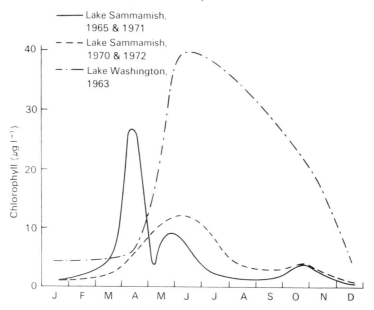

inhibition occurs. This relationship, if quantified, could be expected to hold for one species, one temperature, and at one level of adaptation.

Light intensity can be measured in several units. Flux or footcandles are measures of intensity. Flux is measured as g cal cm^{-2} per minute

Figure 49. Relative penetration of different wavelengths of light through lake water.

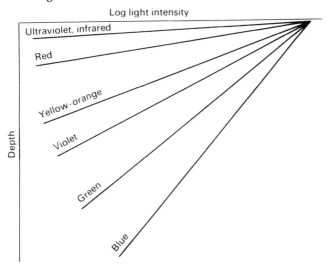

Figure 50. Absorption spectrum for chlorophyll compared with the penetration in water and emission from sunlight. Letters indicate colors. (Schiff 1964)

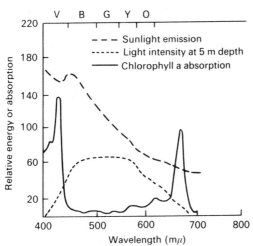

(langleys per minute). Talling (1957, 1965) suggested that light saturation occurs at about 1.7 g cal cm^{-2} per hour (or about 4 800 lux) and inhibition at an intensity of about 8.6–12.9 g cal cm^{-2} per hour (or 24 500–37 800 lux). The bounds for light saturation are thus about 6 and 30 to 50% of full sunlight (78 000 lux, 0.45 g cal cm^{-2} per minute).

The attentuation of light through lake water is an important concept. Light is absorbed in a lake water column according to the following

$$I_Z = I_0 e^{-KZ}$$

where I_Z is the intensity at any depth, I_0 is incident light, Z is depth, and K is the extinction coefficient, which is very useful in describing different water bodies or seasonal changes in one water. Thus a typical light-depth relationship is shown below:

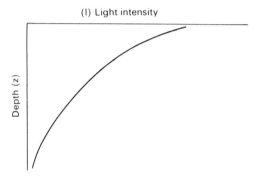

Photosynthesis with depth

Considering light attentuation with depth and the light-photosynthesis relationship, it is entirely understandable that the following

Figure 51. Hypothetical relationship between photosynthesis and light intensity.

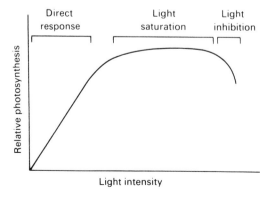

profile results when photosynthesis is measured in the water column (Figure 52).

Light intensity may be so great at the surface that photosynthetic rate is inhibited. At some greater depth the optimum occurs or the photosynthetic rate is saturated at that temperature. Photosynthesis decreases exponentially from that point in proportion to light and usually has decreased to insignificant levels at about 1% of surface intensity, which defines the photic-zone depth. In many clear lakes significant photosynthesis extends below 1% intensity. The compensation depth occurs where respiration equals photosynthesis and no growth results below that depth. This model applies only to a well-mixed photic zone, which unfortunately is not a typical situation in lakes, although it is more so in the ocean. Complete mixing is an acceptable assumption for explaining the process, but it must be remembered that the stratification of nutrients and algal biomass, which occurs in most lakes, will affect the photosynthetic profile. This process, however, does occur to the extent that light is the controlling factor, and, consequently, these kinds of profiles will usually be observed.

This pattern was evident in Chester Morse Lake, an oligotrophic lake in western Washington (Figure 53). Other example profiles of photosynthesis and the significance of integrating over a great depth are shown in Figure 54 for a sewage pond (photic zone 1.5 m) and the Sargasso Sea (photic zone 100 m). Although the total production is not

Figure 52. Relative photosynthesis and respiration versus lake depth, showing varoius characteristics of the water column.

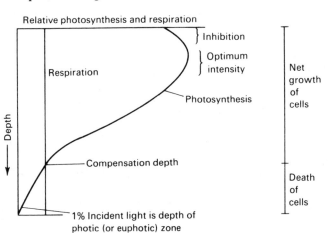

greatly different, the concentration of organisms is obviously at least several orders of magnitude greater in the sewage lagoon.

Efficiency

Only a small percent of available light is actually used because there are too few algae and too much light. About 1% of total incident light is fixed in photosynthesis as a biosphere mean, or about 2% of the visible light because the visible fraction represents about one-half the total light energy. In water this efficiency drops below 1%; Riley reported 0.18% for an oceanic mean (Odum 1959).

Efficiency can be increased greatly by decreasing the light intensity and maximizing the cells' exposure to it. Cultures of *Scenedesmus* and *Chlorella* have utilized as much as 50% of the incident light under these

Figure 53. Distribution of productivity (relative) with depth in Chester Morse Lake, Washington. Values are based on means over a 1.5-yr period. (After Hendrey 1973)

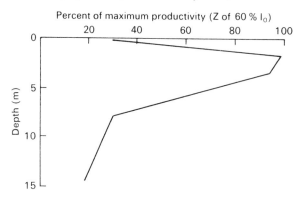

Figure 54. Comparison of productivity profiles in a sewage lagoon and the Sargasso Sea. (Modified from Ryther 1960)

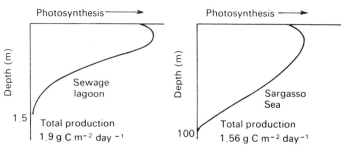

conditions (Brock 1970). The cell itself can increase efficiency by adjusting its chlorophyll content – shade-adapted cells tend to contain more chlorophyll than light-adapted cells. For example, in Chester Morse Lake a photosynthetic efficiency index ($PB^{-1}I^{-1}$) was shown to increase greatly with depth, whereas chl a increased only slightly (Figure 55).

Prediction of photosynthesis and blooms

Total photosynthesis in the case of a well-mixed photic zone can be generally considered to be a function of mean light intensity. Within the mixed layer, the light intensity at the midpoint represents the mean light that a mixed plankton cell will receive. That light level can be defined as:

$$\bar{I} = I_0 \frac{1 - e^{-KZ}}{KZ}$$

where Z is the mixed-layer depth. As an example, $\bar{I}$ would be 20 and 43% for K values of 0.5 and 0.2, respectively. Because this value incorporates the factors of extinction coefficient and depth of mixing, it would be infinitely more practical to use than surface intensity in attempting to predict productivity.

Figure 55. Distribution of chlorophyll a and normalized productivity in Chester Morse Lake. Values are means over a 1.5-yr period. (After Hendrey 1973)

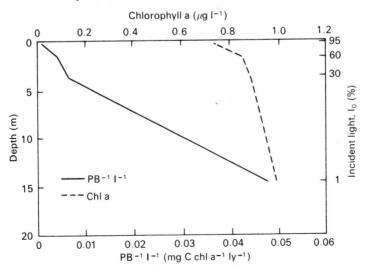

Although productivity cannot be accurately predicted from surface intensity without considering extinction coefficient and mixing depth, there may yet be little change with $\bar{I}$ because of adaptation and inhibition. Inhibition is included in the following equation for growth rate (modified from Steele 1962):

$$\mu = \mu_m \frac{R}{R_0} \exp\left(1 - \frac{R}{R_0}\right)$$

where R is light intensity and r_0 is the optimum or saturated intensity. A curve fitted to data is shown in Figure 56. The rather low optimum intensity is a result of adaptation of natural populations to low laboratory light conditions.

When the aspect of inhibition is applied to a hypothetical situation, one can see how total photosynthesis per unit lake surface could vary little with a sizable change in $\bar{I}$. With mixing depth constant at 8 m and K at 0.3, a decrease in I_0 from 100% to 50% would result in a decrease in $\bar{I}$ from 38% to 19%. Assuming no adaptation and R_0 remains constant at 40% there clearly would not be as great a difference in total photosynthesis as implied by the factor of 2 in $\bar{I}$ (Figure 57). In fact, the greater $\bar{I}$ would produce only 16% more than the lesser $\bar{I}$ value, not 100% as implied. Of course, as R_0 increased, the difference would become greater.

The timing of the vernal outburst can be predicted with knowledge of three variables: (1) light intensity at the surface, (2) compensation depth, and (3) the mixed-layer depth (depth of the thermocline if the epilimnion is well mixed). To understand how these factors can control

Figure 56. Relationship between optimum light intensity (R₀) and growth rate for a natural phytoplankton community. (After Hendrey 1973)

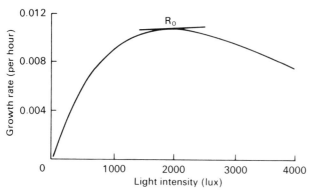

the timing of the outburst, the concept of "critical depth" as defined by Cushing (Raymont 1963) must be understood.

Figure 58 shows an increasing compensation depth as light intensity increases in the spring. "Critical" depth is that depth at which the total photosynthesis (gross) is equivalent to total respiration per unit of surface (or when net photosynthesis is zero). If plankton are distributed evenly with depth, photosynthesis decreases exponentially coincident with light intensity, but respiration is assumed constant with depth. This is probably close to being true in a well-mixed situation. For a phytoplankton bloom to occur, total photosynthesis must exceed respiration and this occurs only when the mixed layer is at less than the "critical" depth.

From surface to the critical depth, production exceeds respiration and net growth can occur. The compensation depth is also the depth where $P = R$ except that it refers to a stationary plankton cell. In reality,

Figure 57. Relative light and growth rate if R_o and K are constant at 40% and 0.3, respectively.

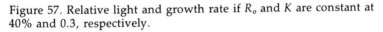

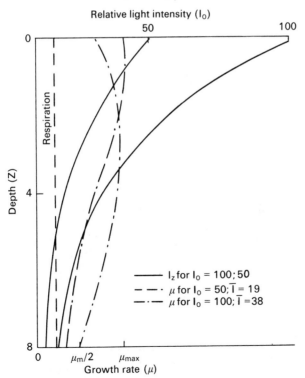

however, the cells are swept below this depth for a considerable period of time and by the same token are swept up near the surface for some period. The net effect is that if the average light intensity received by a mixed plankton cell is greater than the compensation-depth light intensity, net production will result. In many clear-water cases, such as the ocean, this average light intensity can be received if cells are mixed to depths not greater than 5 to 10 times the compensation depth. In many clear lakes the compensation depth may be from 50 to 100 m. Obviously, few lakes are that deep over much of their area, so in spring the mixing depth may be, in effect, the lake bottom, and as light intensity increases, the critical depth may be greater than the maximum lake depth as early as February and an outburst can and does occur. In other situations, such as when turbidity is high, the decreasing mixing depth as thermal stratification increases can determine the bloom timing. This subject will be discussed again when the negative and positive effects of stratification are considered.

Although the critical depth concept developed for oceanic phytoplankton may not be entirely appropriate for lakes, because they are shallower than Z_{cr}, the ratio of $Z_{eu}:Z_m$ (euphotic:mixing) is nonetheless an important concept for lakes (Talling 1971). This can be demonstrated with the following hypothetical example. Given two morphologically different lakes with the same mixing and critical depths, the lake with the greatest ratio of $Z_{eu}:Z_m$ will have the greatest productivity, all other things being equal (Figure 59). To better illustrate

Figure 58. Illustration of the critical depth concept. (Modified from Marshall 1958)

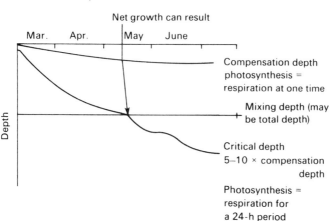

the morphometric differences, use of mean depths of the euphotic and mixing depths would be more appropriate. Thus, for mixing depth 1 the potential productivity ratio would be about 1.0:

$$\frac{\text{Productivity lake A}}{\text{Productivity lake B}} = \frac{\bar{Z}_{eu}/\bar{Z}_{m1}}{\bar{Z}_{eu}/\bar{Z}_{m1}} = \frac{0.5/1.0}{0.4/0.8} \simeq 1.0$$

However, with a doubling of the mixing depth, productivity in B should exceed that in A because of the increased shallow area in B.

$$\frac{\text{Productivity lake A}}{\text{Productivity lake B}} = \frac{\bar{Z}_{eu}/\bar{Z}_{m2}}{\bar{Z}_{eu}/\bar{Z}_{m2}} = \frac{0.5/2.0}{0.4/1.3} \simeq 0.8$$

and therefore as mixing depth increases in the two lakes, the time spent by plankton in the photic zone will be relatively longer in lake B than in lake A. Thus, although the concept of Z_{cr} is not strictly applied in lakes, it can help to explain why some lakes are more productive than others, notwithstanding similar enrichment levels.

Another important light-related factor that limits plankton growth is the attenuation caused by the plankton themselves. As the concentration of plankton increases during a bloom, the light scattered and absorbed by the cells increases and the algae limit their own photosynthesis. This is termed *self-absorption*. Because of self-absorption, photosynthesis is usually diminished at chl a densities greater than about 250 mg m^{-2} (Wetzel 1975, p. 337).

Temperature effects on water-column stability and productivity

The strength and depth of mixing are determined by the degree of stratification, which, as we have seen before, determines the average amount of light available to a mixed plankton cell. If stratification

Figure 59. Two hypothetical lakes with different morphometry but similar euphotic, mixing, and critical depths. The dotted ovals represent mixing patterns for plankton under the two mixing depths.

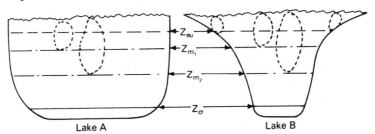

Lake A   Lake B

persists, nutrients become depleted and production is slowed in spite of favorable light. Examples of stable and unstable water columns are indicated in Figure 60 for a hypothetical freshwater lake and a large estuary (Puget Sound). The factor σ_t is defined as the density of water resulting from salinity, temperature, and pressure minus 1 multiplied by 10^3. A common way of indicating the degree of stratification is the change in density per unit depth. In freshwater, density is usually a function of temperature solely, as discussed in Chapter 3.

The positive and negative effects of stability on plankton production are illustrated by the following examples.

Positive effects

In the Indian Arm, a fjord area in British Columbia, productivity was related to the degree of stability and to the compensation depth, which of course increased with time as incident light increased. When these two factors, which determine the amount of light received by a mixed plant cell, were greatest, productivity was greatest (Figure 61). The periods of low stability were favorable to nutrient replenishment.

It is useful to compare the degree of stability in Indian Arm with that in Puget Sound. The unit $(10^5 \mathrm{m}^{-1})$ is actually $(\rho - 1) \times 10^5 \mathrm{m}^{-1}$. The values in Indian Arm ranged from 40 to 100 over the first 10 m, whereas in Puget Sound productivity was strongly affected by stability factors ranging from only 2 to 6 over 50 m depth (Winter et al. 1975). Peaks in productivity and biomass usually occurred at minimum tidal-prism thickness (amplitude), which in turn allowed maximum stability and light availability during the spring months (Figure 62).

In the Duwamish estuary at Seattle, there is no real phytoplankton activity until August when river flow is low and tidal flushing action

Figure 60. Examples of stable and unstable water columns. Dotted lines indicate mixing in the freshwater lake. T, S, and σ_t are, respectively, temperature, salinity, and density for Puget Sound.

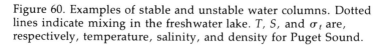

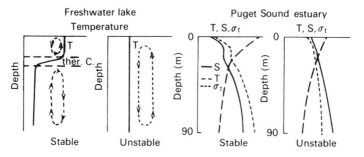

Figure 61. Seasonal relation among compensation depth, water column stability, and net primary production in a British Columbia fjord. (Modified from Gilmartin 1964)

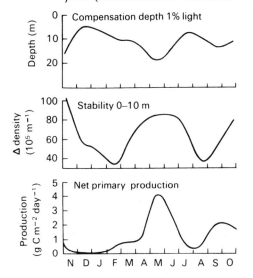

Figure 62. Productivity, biomass, chlorophyll a, stability, and tidal prism thickness in Puget Sound, 1966 (Winter et al. 1975)

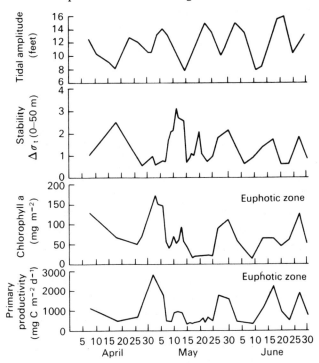

is least. Such tidal conditions are characterized by low-high and high-low tides. The tidal conditions determine the turbulence and stability and are indicated, in addition to river flow, by "tidal-prism thickness," which can be calculated as follows (units in meters):

Unfavorable turbulence: TPT = (HH + LH) − (HL + LL)

$$(3.6 + 3.0) \ − (1.5 + 0.9)$$

$$= 6.0$$

Favorable stability: TPT = (3.0 + 2.7) − (1.8 + 0.9) = 3.0

Figure 63 shows the timing of phytoplankton blooms at one point in the Duwamish related to freshwater discharge and tidal-prism thickness in 1965 and 1966. The water column reaches maximum stability and minimum turbulence and flushing when tidal-prism thickness (difference between sum of two daily high tides and sum of two daily low tides) and freshwater discharge are least. At that time tidal excursions are also minimal and the net result is that surface water moves slowly up and down the estuary and is allowed to retain heat and become even more stable because of the increased temperature gradient. Production correlated with degree of stratification (Δtemperature from surface to bottom) and river discharge, and these two variables explained 65% of the variation in that parameter (Welch 1969a).

Another way to examine the relationship between the physical limitation and productivity in the Duwamish is with a simple growth model. The change in biomass X is a result of growth and loss:

$$\frac{dX}{dt} = \mu X - DX$$

where μX is growth and DX is loss through washout. Because nutrient content is very high and nonlimiting (Welch 1969a), the observed change in mixing depth can be hypothesized to affect the daily maximum growth rate based on ^{14}C productivity measurements made during the bloom. Thus, during neap-tide conditions (TPT = 3 m), the mixing depth was only 1 m and the average μ for a plankton cell mixed to that depth and average light exposure was about 0.6 per day. On the other hand, the spring-tide situation usually resulted in a mixing depth of 4 m and about one-half the μ (0.3 per day) because of the much reduced light per cell. Extinction coefficient was not considered in this case, but higher values usually occur with greater river discharge, which would cause further limitation during the spring runoff period.

The loss rate through flushing can be approximated from a dye study conducted in the estuary, which showed an average flushing (dilution) rate of about 0.2 per day over neap and spring tides alike. Knowing that spring tides allow more plankton loss than neap tides, the flushing rate was adjusted linearly so that a loss rate of 0.13 per day was assigned to the neap tides and 0.27 per day to the spring tides. Even so, it is clear from these values that mixing depth produces a much larger effect on population growth than flushing rate on loss during the change

Figure 63. Tidal prism thickness, river discharge, and chlorophyll *a* content in the Duwamish River estuary, Washington, 1965 and 1966. (Welch 1969a)

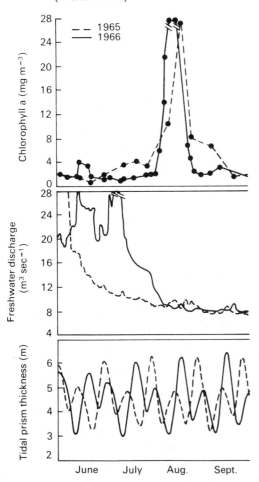

from spring to neap tides. That is a 0.3 per day greater growth, but only a 0.13 per day lesser loss.

The respective net rates of population increase for the neap and spring tide situations, respectively, are:

$$\frac{dX}{dt} = 0.6X - 0.13X = 0.47X$$

and

$$\frac{dX}{dt} = 0.3X - 0.27X = 0.03X$$

Clearly a bloom is not apt to occur in the Duwamish when spring tidal conditions persist, even at low river flow, and this is precisely what has been observed. However, if the μ of 0.47 per day and an initial biomass of 3 μg l^{-1} chl a are used to simulate the bloom initiation of 1966 (Figure 62), the bloom seems explicable. The time required for the biomass to reach the observed maximum of 70 μg l^{-1} chl a is 6.7 days from the first order growth equation:

$$70 \ \mu g \ l^{-1} = 3 \ \mu g \ l^{-1} \ e^{(0.47)(6.7)}$$

Although the dynamics of the phytoplankton in that estuary are, of course, more complex, this analysis nevertheless indicates how strong the effect is from physical factors alone, particularly as they affect the amount of light received.

Negative effects

Stratification can limit production, however, if the condition is prolonged, because of nutrient depletion in the epilimnion. Partial destratification during midsummer in a Thames River reservoir illustrates the part played in production and species succession by stratification and nutrient depletion.

The typical plankton succession occurs during spring and summer, starting with diatoms dominating in the spring, greens and yellow-greens in the summer, blue-greens in late summer, and a small diatom bloom again in late autumn. Taylor (1966) has shown this for a Thames River reservoir (Figure 64).

In mid-July a partial destratification eliminated the typical midsummer growth of green algae and stimulated *Asterionella formosa*, a diatom that usually blooms from February to March. The apparent cause for that summer bloom was the transfer of water at 9–12.5 m with 2.8 mg l^{-1} SiO_2 to the photic zone that contained only 0.5 mg l^{-1} SiO_2. Concentrations below that have been shown to be limiting to *Asterionella* (see section on nutrient limitation).

The *Asterionella* bloom subsequently declined after a second destratification and transfer of fertile water that further increased nutrient levels. Blue-greens continued to increase even during destratification. Population increase was more rapid for *A. formosa* in summer than in spring, but its persistence was much less, possibly a result of temperature that was higher than required for best population maintenance even though growth rate was higher. The concept of thermal optima will be discussed later. Nevertheless, *A. formosa* actually shows greater growth in summer if nutrients are available, because of higher light-saturated photosynthetic rates.

Thus the thermoclines in stratified lakes are clearly a barrier to the recycling of hypolimnetic nutrients and they can have a considerable range of negative effects on plankton algal production, depending on the permanency of the thermocline (see Chapter 6 for further discussion).

Temperature effects on phytoplankton growth rate

The metabolic response of all organisms follows a general law of doubling with each 10 °C increase – the Q_{10} law. This is approximated in most phytoplankton studies of phytopsynthesis and respiration if other factors are not severely limiting. As seen in Figure 65, temperature limits the light-saturated rate of photosynthesis. At low light intensities photosynthesis increases in proportion to light but reaches some maximum that depends upon temperature – as temperature is increased these maxima increase according to the Q_{10} law.

Figure 64. Succession among diatoms, green algae, and blue-green algae in a Thames River reservoir in years with and without destratification. Arrows indicate time of destratification. (Taylor 1966)

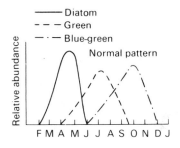
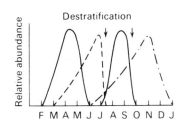

Growth rate

Eppley (1972), who has summarized pertinent data regarding the growth rate of algae related to temperature, has suggested that the maximum rate (μ_{max}) varies with temperature as shown in Figure 66. This relationship includes a variety of species. The Q_{10} from this relationship was shown by Goldman and Carpenter (1974) to be 1.88,

Figure 65. Relationship of photosynthesis of *Chlorella* with light and temperature.

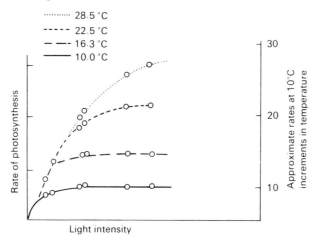

Figure 66. Relation of growth rate (μ_{max}) to temperature for a variety of culture experiments with different species (points). (Eppley 1972)

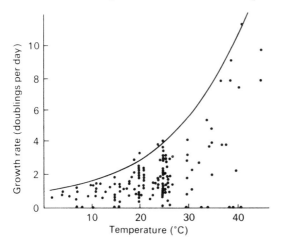

whereas other results from an analysis of a variety of species showed a Q_{10} of slightly more than 2. Nevertheless, even over a broad range of species and temperature optima, the general Q_{10} law holds true with phytoplankton.

Within a curve such as this are included curves that describe the optimum temperature range for growth of a variety of species, each with its one thermal optimum. Four such curves are shown in Figure 67.

Seasonal succession

With such a distribution of thermal optima within a phytoplankton community, changes in temperature alone from season to season should force succession among species according to those optima. If optima for light are also considered, a further dimension is added to account for succession. For example, the following pattern might be expected (Hutchinson 1967):

Winter: Low light and low temperature (no growth)
Spring: High light and low temperature (diatoms)
Summer: High light and high temperature (greens)
Autumn: Low light and high temperature (blue-greens)

Figure 67. Maximum growth rate (μ_{max}) versus temperature with four species of phytoplankton. (Modified from Eppley 1972)

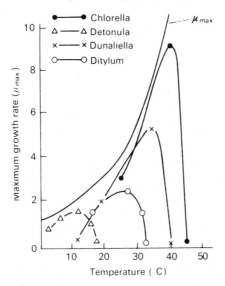

The observed patterns of plankton algae succession are to some extent explainable by light and temperature changes. For example, Cairns (1956) has shown that algae in stream water from Darby Creek in Pennsylvania demonstrated the following shifts in response to raised and lowered temperature from ambient after sufficient time was allowed for population equilibration at each temperature:

$$20\,°C \rightleftarrows 30\,°C \rightleftarrows 35\text{--}40\,°C$$

diatoms   green algae   blue-green algae

Adequate evidence exists that plankton algae do have readily definable thermal optima as shown by Rodhe (1948). He also showed that if such thermal optima are designated by that temperature at which population growth is most prolonged, the temperature is lower than if the criterion of rapid growth is chosen. For *Melosira islandica* Rodhe found that the most prolonged growth occurred at 5 °C, growth was less prolonged at 10 °C, and at 20 °C growth was fast but death was also rapid and the population did not persist. Typical optima are as follows:

> *Melosira islandica* (diatom) $\sim$ 5 °C
> *Synura uvella* (yellow-green) $\sim$ 5 °C
> *Asterionella formosa* (diatom) 10–20 °C $\sim$ 15 °C
> *Fragilaria crotonensis* $\sim$ 15 °C
> *Ankistrodesmus falcatus* $\sim$ 25 °C
> *Scenedesmus, Chlorella, Pediastrum, Coelenastrum* 20–25 °C

A typical pattern of plankton algae succession is shown in Lake Sammamish (Figure 68). Note the spring diatom outburst and the summer and autumn blue-green dominance. The diatom increase was greatest at temperatures between 5 and 10 °C, whereas blue-greens increased when temperature was between about 15 and 20 °C. *Fragilaria* and *Asterionella* were important representatives in the diatom outburst.

An explanation for thermal optima being lower than the temperature for maximum growth rate is that as μ increases exponentially with increasing temperature, the death rate must also increase, but probably lags behind μ. Because biomass would be the difference between growth and death, its maximum would necessarily occur at a lower temperature than for μ. This can be hypothetically illustrated (Figure 69).

There are many exceptions to light and temperature as an explanation for such successional patterns in phytoplankton. *Oscillatoria rubescens* has occurred in Lake Washington at temperatures greater than 20 °C, and was the principal bloom species that first indicated the degradation in that lake, yet it prefers low light and low temperature, around 10 °C. *Fragilaria crotonensis* should occur most prominently in spring and autumn because its optimum is from 12 to 15 °C, yet its maximum in

Figure 68. Surface phytoplankton composition at a centrally located station in Lake Sammamish. Temperature indicated by solid circles. (After Isaac et al. 1966)

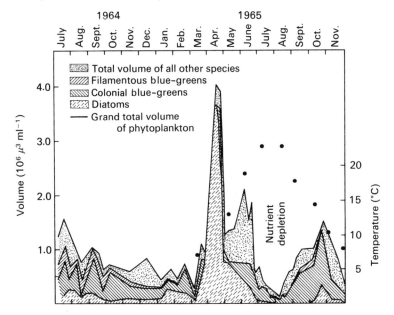

Figure 69. Hypothetical relation among growth rate, death rate, and population mass and temperature for a species of algae. Note the optimum (vertical line) for biomass is lower than for growth rate.

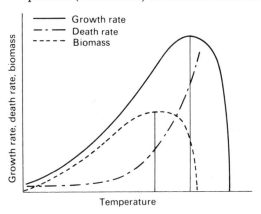

nature is from 15–29 °C. Moses Lake produces massive blooms of blue-green algae of the genera *Anabaena*, *Aphanizomenon*, and *Microcystis* (often referred to as "Anny," "Fanny," and "Mike") at an average summer temperature of 22 °C (Bush et al. 1972). Clear Lake, California, shows the same situation of blue-green blooms occurring during moderate temperatures. Results are shown in Figure 70 for a calm autumn day during which *Aphanizomenon* was the dominant species (Goldman and Wetzel 1963).

The occurrence of blue-green algae tends to be initially confined to late summer or autumn in moderately enriched lakes, but as enrichment proceeds, they begin to dominate even in spring, which is the case in early spring in Pine Lake, Washington, at a temperature of 8 °C.

Why are light and temperature and their respective optima not adequate in predicting successional patterns within and among lakes?

Species adaptation

Algae are capable of adaptation to other temperature regimes even though somewhat removed from their optima. For example, the respiration rates of *Chlorella* show that an adaptation of strains can occur (Table 8). Although one strain obviously does best in the upper 30s, as was indicated in Figure 67, another strain nevertheless can be rather active at lower temperatures.

Blooms of *Oscillatoria* can be observed below the ice during the dead of winter in lakes. Of course, their growth rate is low at that time, just

Figure 70. Observations of productivity and temperature, Clear Lake, California, on October 15, 1959 (a calm day). *Aphanizomenon* is the dominant species. (Modified from Goldman and Wetzel 1963)

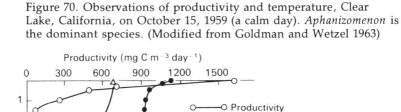
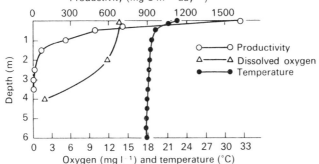

as is the rate of *Chlorella* at less than optimum temperature (Table 8). However, if loss rates are also low, a bloom can result.

Thus, algae should be able to compete at temperatures slightly removed from their range and possibly dominate the community if some other factor(s) is (are) provided nearer their optimum and to some extent compensate the organism for the out-of-optimum temperature.

Interaction with other factors

If provided closer to a species optimum, nutrient supply may be one such factor that may allow that species to grow outside its temperature optimum. If the supply is great enough, even if the growth rate is relatively low, a bloom can result. This may, for example, explain blue-green algae blooms (*Aphanizomenon*) at 8 °C in spring at relatively high light intensities. Even if temperature is suboptimum, blue-greens may be able to outcompete diatoms and greens if the nutrient supply is more to their liking. This may hold for *O. rubescens*, which adapted to the warm season in Lake Washington because the nutrient supply was increased. Such selection for higher P levels and low light is indicated in Table 9 for filamentous blue-greens.

The flotation mechanism allows blue-greens to compete with more dense diatoms for light in warm, quiescent conditions (Figure 70). Diatoms depend on turbulence to remain suspended in the lighted zone. Thus, with the lower sinking rate, the blue-greens are able to accumulate large masses when turbulence is minimal, assuming the nutrient supply is large.

Some blue-greens are known to impart toxic excretory products in water, thus causing deaths in domestic cattle. *Microcystis* has been shown to grow best in water from which it was filtered, apparently "conditioning" the water for its own growth (Vance 1965), whereas

Table 8. *Effect of acclimation temperature on respiration in two strains of* Chlorella pyrenidosa

| Test temp. | Respiration[a] for two acclimation temp. | |
	25 °C	38 °C
25 °C	4.5	8
39 °C	1.6	18

[a] $mm^3 O_2$ respired mm^{-3} cells hr^{-1}
Source: Sorokin (1959).

other forms did not grow in such water. Keating (1977) has convincingly shown that the extent of blue-green dominance in the summer in Linsley Pond was do in large measure to excretory products. Blue-green abundance in summer, relative to diatoms, was related to blue-green abundance in the previous winter. The inhibition was also demonstrated in culture. Although their relative importance is generally unclear, such antibiotic factors no doubt play a role in succession and could explain to some extent why blue-greens bloom at temperatures below their optima.

Summary of effects of temperature and light

Although temperature and light contribute to species dominance and succession, they do not tell the whole story because the

Table 9. *Variation in the relative abundance of greens, blue-greens, and diatoms resulting from five levels of phosphorus enrichment at three light intensities in Lake Washington water*

μg Pl^{-1} added	Light intensity in lux			Row means
	4000	2000	1000	
Greens (Number of cells)				
0	43	58	32	44
10	132	196	57	128
20	211	143	65	140
30	243	285	86	204
40	334	232	92	219
Column means	193	183	66	
Blue-greens (μ of filament length)				
0	1220	1749	2680	1883
10	1860	4540	4160	3520
20	3120	5010	5620	4583
30	1840	3980	6000	3940
40	2840	6960	5440	5080
Column means	2176	4448	4780	
Diatoms (Number of cells)				
0	5	2	9	5
10	153	69	30	84
20	91	35	24	50
30	168	86	15	90
40	135	30	13	59
Column means	110	44	18	

Source: Hendrey (1973).

available nutrient supply and other factors are also elements in the adaptation of species to sub- or supra-optimum temperature and light. The blue-green species that usually prefer warm summer temperatures can apparently outcompete other algae for the available nutrients at that time. However, they can also adapt to other seasons when nutrient availability is high. In spite of suboptimum temperature, blue-green species can dominate for most of the year in overfertilized lakes.

Although a given temperature is not entirely indicative of species dominance among different waters, succession from diatoms to greens to blue-greens is often reasonably predictable from a temperature increase imposed upon a given water (and nutrient supply) and temperature regime.

The interacting physiological effects of temperature and light on production can produce two results:

1. A short-term effect of a temperature increase may result in some adaptation of existing species, with the possible occurrence of even a decrease in production if the original temperature optimum and the new temperature are too far apart.

2. A long-term species shift may occur with time, and species with optima at the new temperature will dominate and production will increase because the light-saturated rate of photosynthesis goes up with a temperature increase across the tolerable range for all species.

Heated waters from power plants into lakes

Not many examples exist to examine the effect of heated-water addition, but results from some Polish lakes indicate that fairly extensive changes in productivity and species composition could develop from relatively small changes in temperature (Hawkes 1969):

Lichen Lake (receives heated waters from a power plant): test lake
Annual temperature range: 7.4–27.5 °C
Taxonomic components: 285 (sp., varieties, and forms)
Dominant species: *Melosira amgibua, Microcystis aeruginosa*
Primary production: 7.3 g m^{-2} day^{-1}

Slesin Lake (no heated water added): control lake
Annual temperature range: 0.8–20.7 °C
Taxonomic components: 198
Dominant species: *Stephanodiscus astraea* (no blue-greens)
Primary production: 3.75 g m^{-2} day^{-1}

Eutrophication

Definition

This process can be defined as an increase in enrichment that causes increased productivity in lakes, streams, and estuaries. Increased enrichment can occur from an increase in the external nutrient supply rate (Beeton and Edmondson 1972). This increase can come from natural processes such as earthquakes, forest fires, and so on, and thus result in periods of high enrichment during the life of a lake. The lake and watershed could stabilize following such a catastrophe and fall back to earlier levels of enrichment.

When the increased external nutrient supply is man-made, it is termed *cultural eutrophication*. Sources of such nutrients include domestic sewage, rural runoff, fertilizer, urban (storm) runoff, and detergents.

Increased enrichment can also result after thousands of years as a result of decreased lake volume through sedimentation and filling in. This best describes the "natural aging process" often attributed to lakes. Even without an increase in the external nutrient supply rate, a lake can become more enriched because the decrease in lake volume results in an increased concentration of nutrients. Although the decreased volume results in a reduced residence time for incoming nutrients, the decreased volume also increases the sediment surface:lake volume ratio, which increases the importance of the sediment as an internal source of nutrients.

The sediment that causes lake filling is organic or inorganic and originates from autochthonous or allochthonous sources. Sedimentation rates are often on the order of 3 mm per year (the rate in Lake Sammamish and Lake Washington). At such a slow rate of sedimentation, increased enrichment through the natural aging process (decrease in volume) would necessarily require a long time under normal conditions of erosion. Therefore, if nutrient concentrations and productivity show significant increases in the past 50 to 100 years in a lake, the enrichment is probably man-caused, barring a natural catastrophe, and no doubt is cultural eutrophication.

Rate of eutrophication

The rate at which eutrophication proceeds in a lake depends upon the following factors if the nutrient loading is constant.

Trophic state of a lake. Lakes that are hypereutrophic, or highly enriched, probably do not show as much response to a given increase in nutrient supply as an oligotrophic or mesotrophic lake because the effect of further enrichment diminishes when light intensity begins to limit production as a result of self-shading from a large plankton biomass. On the other hand, oligotrophic lakes may show a slower response than mesotrophic or mesotrophic-eutrophic lakes because the sediments would act as a more permanent sink for phosphorus.

Mean depth of a lake. This factor is important because, as an index of lake volume, it corrects for dilution and determines the resulting concentration from a given input of nutrients. A decrease in mean depth results in more available light for production and thus potential for nutrient use. The sediments are closer to the photic zone and epilimnion and also allow for more reuse of the nutrient income.

The significance of nutrient supply and lake depth is shown in Figure 71. In this example from the Precambrian shield area of Canada, Schindler (1971a) assumed that output from the watershed to the lakes had achieved a long-term steady state with input from rainfall. Thus the nutrient supply per volume (v) was considered proportional to lake area (A_0) and watershed area (A_d) multiplied by $1/v$. As volume of the lake decreased or as watershed area increased, lake response in terms of productivity would also be expected to increase because area : volume ratio is an index of the resulting steady-state nutrient concentration in the lake. The results are for several pristine lakes in the area; lake response clearly showed a good relation with the lake-nutrient-concentration index.

Figure 71. Relationships of lake response to watershed (A_d) plus lake (A_0) area divided by volume (v). (Schindler 1971a)

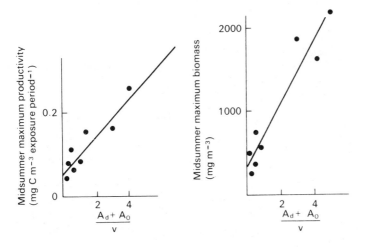

Morphometry. Lakes with small hypolimnia offer more potential for nutrient reuse and available light than lakes with large hypolimnia. The smaller the hypolimnion, the less total oxygen it contains and the sooner it will reach anaerobiosis.

Flushing. The greater the flushing rate, the shorter the period of time available for use and reuse of the nutrient income. And if the flushing rate is high enough, plankton buildup may be prevented because of cell washout.

The effect of flushing has been best demonstrated by Dillon (1975) for Cameron and Four Mile lakes in Ontario. Although these two lakes had approximately the same trophic state, the flushing rate of Cameron was 70 times that of Four Mile Lake, whereas the P load to Cameron was 20 times that to Four Mile. Correction of loading for flushing accurately placed these two at the same relative state.

Characteristics related to lake trophic state

Some characteristics often observed in lakes follow various gradations in quantity and quality from oligotrophy to eutrophy and can be used to judge the relative stage in a lake's development. From a qualitative standpoint such characteristics are listed in Table 10.

Table 10. *Qualitative characteristics of oligotrophic and mesotrophic lakes*

	Oligotrophic	Eutrophic
Depth	Deep	Shallow
Hypolimnion: epilimnion	> 1	< 1
Primary productivity	Low	High
Rooted macrophytes	Few	Abundant
Density of plankton algae	Low	High
Number of plankton algal species	Many	Few
Frequency of plankton blooms	Rare	Common
Depletion of hypolimnetic oxygen	No	Yes
Fish species	Cold water, slow growth, restricted to hypolimnion	Warm water, fast growth, tolerate low O_2 in hypolimnion and high temperature of epilimnion
Nutrient supply	Low	High

Criteria for the trophic state of lakes

Considerable effort has gone into quantitative definition of lake trophic state in recent years. To determine reference trophic states of lakes and to detect changes in trophic states, agreed upon criteria are necessary. A significant increase in any of these listed indices, detectable at most over a 5-year period, could be considered as sufficient evidence that accelerated cultural eutrophication is occurring. A slight change over a shorter period would probably not, then, necessarily indicate a lack of accelerated eutrophication. A change detectable only after 5 years may still indicate unnaturally accelerated eutrophication, but 5 years is suggested as probably a realistic maximum for the average monitoring endeavor.

The dynamic characteristics of lakes, the individuality among lakes, and the desires of people may in some instances produce exceptions to these criteria.

Primary productivity

Ranges in photosynthetic rate, as measured by radioactive-carbon assimilation, that are indicative of trophic states have been suggested by Rodhe (1969). These limits, with appropriate modifications, are given in Table 11. Similar values from two other authors are cited by Grandberg (1973). Although these ranges were determined for temperate lakes, recent comparisons of productivity at various latitudes suggest that a more general application may exist.

Although Table 11 shows quite a large range for eutrophy, Rodhe (1969) and others (Grandberg 1973) have separated eutrophic from hypereutrophic at about 40% of the range (250 g C m^{-2} per year and 1000 mg C m^{-2} per day). Of course, the gap between oligotrophic and eutrophic in Table 11 is considered the mesotrophic state.

Table 11. *Ranges in rates of primary productivity as measured by total carbon uptake attributable to the trophic states of lakes*

Period	Oligotrophic	Eutrophic
Mean daily rates in a growing season, mg C m^{-2} day^{-1}	30–100	300–3000
Total annual rates, g C m^{-2} year^{-1}	7–25	75–700

Source: Modified from Rodhe (1969) with permission of the National Academy of Sciences.

Biomass

Chlorophyll *a* is considered the most specific and versatile measure of algal biomass. The mean summer chlorophyll concentration as determined from epilimnetic water samples collected at least monthly is an index of the trophic state of a lake and can be generally defined as oligotrophic, 0-4 μg chlorophyll *a* l^{-1}; eutrophic, 10-100 μg chlorophyll *a* l^{-1}.

These ranges are suggested after reviewing reported data on chlorophyll concentrations and other indicators of trophic state in several lakes throughout the United States and Canada and by comparing linear relationships between total phosphorus and chlorophyll *a* concentration (Dillon and Rigler 1974a; Jones and Bachmann 1976).

Of considerable interest are data from Lake Washington, which show that during peak enrichment mean summer chlorophyll *a* content rose to about 38 μg l^{-1} and the lake was definitely eutrophic. The post-nutrient diversion summer mean has declined to levels from 5 to 10 μg l^{-1} and the lake is more typically mesotrophic (Edmondson 1970 and 1972). Unenriched and relatively low productive lakes at higher elevations in the Lake Washington drainage basin show mean summer chlorophyll *a* contents of 1 to 2 μg l^{-1} (Hendrey 1973). Moses Lake, which can be considered hypereutrophic, shows a summer mean of 90 μg l^{-1} chlorophyll *a* (Bush et al. 1972). Mean concentrations in the open-water areas would not be expected to greatly exceed 100 μg l^{-1}, because of self-shading, except through wind- or current-caused concentration phenomena.

Two equations relating total P and chl *a* are:

log chl *a* = 1.449 log P − 1.136 (Dillon and Rigler 1974a)

log chl *a* = 1.46 log P − 1.09 (Jones and Bachmann 1976)

Dillon and Rigler used data from 46 lakes and Jones and Bachmann from 143. The former data set included P values from spring turnover and summer chl *a*, whereas the latter was a pool of annual mean values. Because the relationships are log-log, the accuracy of prediction is not great. For the first equation, for example, the errors can be ± 60-170% and 30-40% for 95% and 50% confidence intervals, respectively, for a chl *a* content of 5.6 μg l^{-1}. Nonetheless, the correlation coefficients were high (0.95) for both, so, although there is considerable lake-to-lake variation, the equations apparently represent a definite cause and effect and a basis for a trophic state guideline.

The equations agree very closely. The values of 4 and 10 μg l^{-1} chl *a* suggested as limits for oligotrophy and eutrophy correspond to 16 and 30 μg l^{-1} total P. This point will be reemphasized under nutrient levels.

Oxygen deficit

Criteria for the rate of depletion of hypolimnetic oxygen in relation to trophic state were reported by Mortimer (1941 and 1942) as follows:

Oligotrophic < 250 mg O$_2$ m^{-2} per day

Eutrophic > 550 mg O$_2$ m^{-2} per day

This is the depletion rate of hypolimnetic oxygen determined by the change in mean concentration of hypolimnetic oxygen per unit time multiplied by the mean depth of the hypolimnion. The overall observation time interval should be at least a month, preferably longer, during summer stratification.

Recently, the O$_2$ deficit rate (ODR) has been correlated with areal P loading corrected for flushing rate (ρ) in a representative group of lakes and a correlation coefficient of 0.73 was observed (Welch and Perkins, in press). The equation is:

$$\log \text{ODR (mg O}_2\text{ m}^{-2}\text{ day}^{-1}) = 1.51 + 0.39 \log L/\rho \text{ (mg P m}^{-2})$$

Interestingly, the P loading limit for eutrophy of 400 mg P m^{-2} per year, suggested by Vollenweider (as will be discussed later), and for lakes 10–15 m deep that flush once per year gives values for ODR from the equation of 400 to 433 mg O$_2$ m^{-2} per day. That approximates reasonably well the limit suggested by Mortimer.

By the same token, Edmondson (1966) has related the O$_2$ deficit in Lake Washington with the increase in eutrophication (Figure 72). The O$_2$ deficit in Lake Washington was about 500–800 mg O$_2$ m^{-2} per day, which suggests that the lake was mesotrophic-eutrophic. It also illustrates an anomaly: that is, the ODR did not continue to increase with P loading, which increased greatly between 1955 and 1962. The decrease in 1962 was thought to be the result of more blue-green algae, which tend to float and are decomposed in the epilimnion. The ODR has decreased only gradually in Lake Washington even though the plankton crop has greatly decreased. The answer may be that the accumulated organic matter in the sediments must still account for the ODR (Edmondson, personal communication).

Indicator species

The representation of certain species of freshwater plankton is often a sensitive indicator of trophic state. Rawson (1956) has suggested an approximate distribution of limnetic algae in lakes of western Canada over the range of trophic state:

Oligotrophic   *Asterionella formosa*
                Melosira islandica
                Tabellaria fenestrata
                Tabellaria flocculosa
                Dinobryon divergens
                Fragilaria capucina
                Stephanodiscus nigarae
                Staurastrum ssp.
                Melosira granulata
Mesotrophic   *Fragilaria crotonensis*
                Ceratium hirundinella
                Pediastrum boryanum
                Pediastrum duplex
                Coelosphaerium naegelianum
                Anabaena ssp.
                Aphanizomenon flos-aquae
                Microcystis aeruginosa
Eutrophic   *Microcystis flos-aquae*

Figure 72. Hypolimnetic oxygen deficit below 20 m in Lake Washington in 1933, 1950, 1955, 1957, and 1962. For comparison of rates, constant slopes for deficit rates of 1, 2, and 3 mg O_2 cm^{-2} mo^{-1} are indicated. (Edmondson 1966)

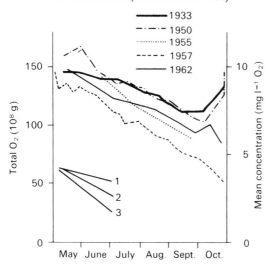

Although Rawson's species list is obviously not appropriate world-wide, it nevertheless offers an example of the type of approach that can be useful in establishing trophic state and change. There is no special order intended in the list of blue-green species. Another such list is provided by Wetzel (1975).

The dominance of the plankton (numerically or in mass) by blue-green algae, such as *Aphanizomenon, Microcystis, Anabaena, Coelosphaerium,* or *Aphanocapsa,* during the warm period represents the most significant phenomenon that inhibits recreational and esthetic use in an enriched water body. The presence of these and their relative abundance and bloom frequency are strong indicators of eutrophy. Unfortunately, there are no quantitative guidelines for any nuisance species that would help identify trophic state.

The presence of blue-green algal blooms is pertinent only in fresh water. Although these organisms also flourish in some enriched brackish waters (Baltic Sea), they do not flourish in marine areas, where the increased biomass caused from overenrichment is due to diatoms and dinoflagellates.

Sediment diatoms

Stockner (1972) has devised a trophic index based on the ratio of Araphidinae: Centrales diatom frustules with depth in the sediment. Essentially, this is a ratio of the pennate to centrate planktonic diatoms and the relationship to trophic state is an empirical one. A ratio of 0–1 indicates oligotrophy, 1–2 mesotrophy, and greater than 2 eutrophy. If an estimate of sedimentation rate is known, a historical pattern of trophic state can be determined. Other sediment data used to indicate trophic change are P, N, and C content; however, no quantitative limits for those variables have been proposed. Nevertheless, changes over time related to sedimentation rates can define much about the trophic history of lakes.

Nutrients – concentration

Critical levels of dissolved inorganic nitrogen and phosphorus at the time of spring overturn in Wisconsin lakes that if exceeded would probably produce nuisance blooms of algae were suggested early on by Sawyer (1947). These concentrations are 10 μg l^{-1} P and 300 μg l^{-1} N. Measured at the spring overturn, prior to the start of the active growing season, nutrient concentrations should be maximum and therefore best represent the supply for plankton algae in the approaching growing season. This is analogous to algal growth in a flask inoculated with nutrients. Nutrient concentrations during active growth periods may only indicate the difference between amounts absorbed in biomass (suspended and settled) and the initial amount biologically available. The values, therefore, would not be indicative of potential algal production. The nutrient form most commonly used to indicate trophic state in lakes is total P. As indicated in the discussion on chl *a*, there is a sound basis for values of 15 and 30 μg l^{-1} total P as limits for oligotrophy and eutrophy because they conform (according to the equations) to the trophic limits for chl *a* of 4 and 10 μg l^{-1}. Because total P is usually 2 to 4 times dissolved P during nongrowth periods, total P in Sawyer's study may have also been close to 30 μg l^{-1}. Others (Vollenweider 1976) consider 10 and 20 μg l^{-1} total P as limits for oligotrophy and eutrophy, respectively.

One of the most convincing relationships between maximum phosphate content at lake overturn and eutrophication, as indicated by algal biomass, has been shown in Lake Washington (Edmondson 1970, 1972). During the years when algal densities progressed to nuisance levels, maximum total soluble P in winter increased from 10–20 μg l^{-1} up to 57 μg l^{-1}. Since sewage diversion, maximum PO_4-P has decreased once again to the preenrichment level. Correlated with the P reduction is chlorophyll, which has decreased from a maximum of 38 μg l^{-1} to a level of less than 10 μg l^{-1}.

Nutrients – loading

Although more difficult to measure than maximum concentration at overturn, the rate of nutrient inflow, or loading, is considered the controllable variable for eutrophication and is more useful for predicting lake trophic state over a broad range of lake types. Loading rates are usually determined by knowing water flow and nutrient concentration in natural surface and groundwater inflows as well as wastewater inflows. There exists an interesting chronology in the concepts of loading rates. Initially, Vollenweider (1968) related nutrient supply to mean depth in several well-known lakes in the world and identified nuisance levels associated with induced eutrophication. From these relationships, shallow lakes were clearly seen to be more sensitive to nutrient income than deep lakes. This is simply because the concentration effect increases as lake depth decreases, creating higher concentrations as well as increased nutrient reuse. From this standpoint, nutrient supply is a much more useful criterion than concentration in predicting trophic state. Examples of critical loading limits are about 0.3 g m^{-2} P per year and 4 g m^{-2} N per year for a lake with a mean depth of 20 m and about 0.8 g m^{-2} P per year and 11 g m^{-2} N per year for a lake with a mean depth of 100 m, the deeper lake having more dilution for assimilating incoming P (Figures 73 and 74). The equation that defines the critical loading for eutrophication is:

$$L_c \ (\text{mg m}^{-2} \ \text{yr}^{-1}) \approx 50 \ \bar{Z}^{0.6}$$

The first refinement of the loading graph involved a correction for water-residence time (Vollenweider and Dillon 1974). The variable on the abscissa became mean depth divided by water-residence time: $\bar{Z}\tau_w^{-1}$. Because the reciprocal of τ_w is flushing rate (ρ), the variable

Figure 73. Loading graph for N in various lakes. Ae, Aegerisee (Switzerland); Ba, Baldeggersee (Switzerland); Bo, Bodensee, Lake Constance (Austria, Germany, Switzerland); E, Lake Erie (U.S.A.); Gr, Greifensee (Switzerland); Ha, Hallwilersee (Switzerland); Lé, Lake Léman, Lake Geneva (France, Switzerland); Mend, Lake Mendota (U.S.A.); Mo, Moses Lake (U.S.A.); Mä, Lake Mälaren (Sweden); Norrv, Lake Norrviken (Sweden); Ont, Lake Ontario (U.S.A.); Pf, Pfäffikersee (Switzerland); Seb, Lake Sebasticook (U.S.A.); Tahoe, Lake Tahoe (U.S.A.); Tü, Türlersee (Switzerland); Vä, Lake Vänern (Sweden); Wash, Lake Washington (U.S.A.); WE, Western Lake Erie (U.S.A.); Zü, Zürichsee (Switzerland). (Vollenweider 1968)

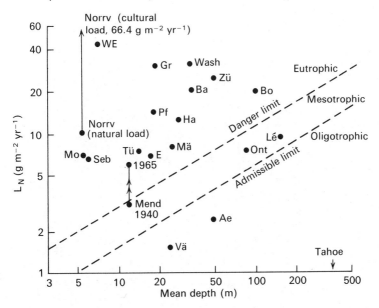

can also be represented by $\bar{Z}_\rho$. Vollenweider (1976) designated this variable as q_s or surface hydraulic loading.

From the initial loading graph, uncorrected for τ_w, lakes like Tahoe that are large and deep and have low loading rates were located far to the lower right-hand corner (Figures 73 and 74). It would appear that Tahoe would never reach a dangerous limit from increasing eutrophication. On the later graph (Figure 75), correction for the large residence time places Tahoe far to the left, showing it to be, in fact, rather sensitive to increased loading. The reverse is true for some lakes orig-

Figure 74. Loading graph for P in various lakes. Ae, Aegerisee (Switzerland); Ba, Baldeggersee (Switzerland); Bo, Bodensee, Lake Constance (Austria, Germany, Switzerland); d'Ann, Lake d'Annecy (France); E, Lake Erie (U.S.A.); Fu, Lake Fureso (Denmark); Gr, Greifensee (Switzerland); Ha, Hallwilersee (Switzerland); Lé, Lake Léman, Lake Geneva (France, Switzerland); Mä, Lake Mälaren (Sweden); Mend, Lake Mendota (U.S.A.); Mo, Moses Lake (U.S.A.); Norrv, Lake Norrviken (Sweden); Ont, Lake Ontario (U.S.A.); Pf, Pfäffikersee (Switzerland); Sam, Lake Sammamish (U.S.A.); Seb, Lake Sebasticook (U.S.A.); Tahoe, Lake Tahoe (U.S.A.); Wa, Lake Washington (U.S.A.); WE, Western Lake Erie (U.S.A.); Vä, Lake Vänern (Sweden); Zü, Zürichsee (Switzerland). For Lakes Erie and Ontario, dotted lines show P loading by 1986 (from 1967 estimates) without P control; dashed lines show P loading by 1986 (from 1967 estimate) with no P in detergents and with 95% P removed from all municipal and industrial wastes. (Vollenweider 1968)

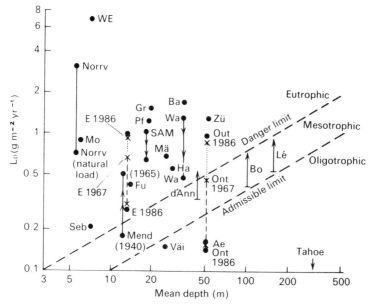

inally located high on the left side of the graph (high flushing and shallow). Those were shifted to the right, a more reasonable position with respect to their observed trophic state. The lakes represented are part of a special study of eutrophication, the North American Project of the OECD (Organization for Economic and Cooperative Development), the results of which were reported by Rast and Lee (1978).

Vollenweider (1976) proposed that the critical loading limit for mesotrophy and eutrophy from that graph were:

$$L_c \ (\text{mg m}^{-2} \text{ yr}^{-1}) \approx 100 - 200 \ (\check{Z} \ \rho)^{0.5}$$

Essentially this model considers the inflow concentration of P and the values of 100 and 200 represent empirical fits for lines separating oligotrophy from eutrophy.

For the most part, the lakes represented in Figure 74 conform to the Vollenweider model in terms of their trophic state. There are a few exceptions, with several mesotrophic and oligotrophic lakes falling above the appropriate line, but such variability would seem to be expected at the transition areas.

Finally, Vollenweider (1976) has refined the relationship even further:

$$L_c = 10 - 20 \ q_s \ [1 + (\check{Z}/q_s)^{0.5}]$$

where now the values 10 and 20 represent steady-state concentrations

Figure 75. U.S. OECD data applied to initial Vollenweider phosphorus loading and mean depth/hydraulic residence time relationship. (Rast and Lee 1978)

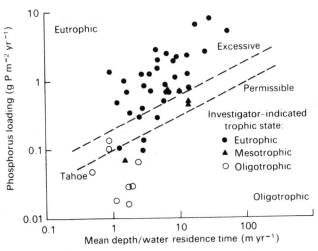

in the lake, because the variable $[1 + (\bar{Z}/q_s)^{0.5}]$ provides a correction for sedimentation or retention of P.

Still another formulation for critical loading has come from Dillon (1975). Dillon transformed Vollenweider's (1969) original model that defines the equilibrium P concentration in the lake:

$$\bar{P} = \frac{L}{\bar{Z}(\rho + \sigma)}$$

where σ is the sedimentation rate coefficient per year. The transformation gives:

$$\bar{P} = \frac{L(1 - R)}{\bar{Z} \, \rho}$$

where R is the retention coefficient or the fraction of P retained in the lake. The limits of equilibrium P proposed for mesotrophy and eutrophy by Dillon are also 10 and 20 μg l^{-1}. If this formulation is graphed and the same NA-OECD lakes are plotted (Figure 76), the results are about the same as for the Vollenweider graph.

Although there is much agreement among the various approaches to defining the critical P loading, there still seems to be some overlap of mesotrophy and eutrophy between 20 and 30 μg l^{-1} P, whether measured or predicted. This seems little discrepancy to live with, considering the myriad of problems and sources of error in estimating nu-

Figure 76. U.S. OECD data applied to Dillon phosphorus loading, phosphorus retention, and mean depth relationship. (Rast and Lee 1978)

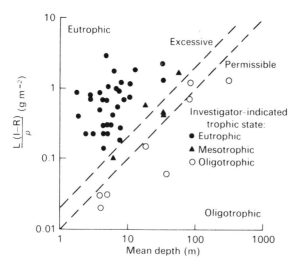

trient loading, trophic state, and the variables included in these equilibrium models.

Several methods for representing trophic state on a quantitative basis have been proposed. However, the most appropriate, because it deals with absolute rather than relative values, is that by Carlson (1977). Using data from a variety of lakes, he has related total P, chl a, and Secchi disk to each other on the basis of $\log_2$. Thus, a doubling in P is related to reduction by half in SD. The respective equations are as follows:

$$\text{TSI} = 10\,(6 - \log_2 \text{SD})$$

$$= 10\left(6 - \log_2 \frac{7.7}{\text{chl } a^{0.68}}\right)$$

$$= 10\left(6 - \log_2 \frac{64.9}{\text{P}_{\text{tot}}}\right)$$

Representative values of P, chl a, and SD, calculated from these equations for appropriate TSI values, are shown in Table 12. The relationship between SD and chl a is particularly important. As represented in Figure 77, the greatest change in SD occurs below a concentration of chl a of about 20 μg l^{-1}. At chl a above 20 μg l^{-1} there is relatively little change in SD. Because lack of water clarity is the result of eutrophication most obvious to the public, it is important for management purposes to realize that chl a must be lowered to levels below 20 μg l^{-1} before much noticeable improvement in water clarify per se can be seen. Other improvements in terms of scums, odors, and the food chain may occur without noticeable improvement in clarity, however.

Table 12. *Representative values of P, chl a, and SD with respective TSIs; values from 0–20 and 70–100 omitted*

TSI	P_{tot}	chl a	SD
20	3	0.34	16
30	6	0.94	8
40	12	2.6	4
50	24	6.4	2
60	48	20.0	1
70	96	56.0	0.5

Source: After Carlson (1977).

Summary of criteria

The primary characteristics that can be used to quantify the trophic state and the associated levels are shown in Table 13. These are total P, chl *a*, ODR, and Secchi-disk depth. As stated previously, these characteristics do interrelate well for the most part. As suggested by the ranges presented, a given lake could be mesoeutrophic or oligo-mesotrophic with respect to some characteristics.

Figure 77. Relationship of Secchi-disk depth to chlorophyll according to equation from Carlson (1978).

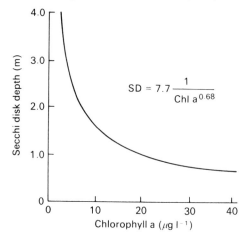

$$SD = 7.7 \frac{1}{Chl\ a^{0.68}}$$

Table 13. *Summary of quantitative limits for several characteristics that define trophic state. For loading and ODR a 10 m deep lake with a 1-year retention time is assumed. Previously given equations were used to estimate the values*

	Oligotrophy $\leq$	Mesotrophy	Eutrophy $\geq$
μg l^{-1} total P (winter)		10–15	20–30
μg l^{-1} chl *a* (summer)		2–4	6–10
m Secchi disk (summer)		5–3	2–1.5
mg m^{-2} yr^{-1} P loading, 10–20 $q_s[1 + (\bar{Z}/q_s)^{0.5}]$		200	400
mg m^{-2} yr^{-1} P loading, 100–200 $(q_s)^{0.5}$		316	632
mg m^{-2} day^{-1} ODR (based on above P loading limits)		250–310	330–400

Nutrients - prediction

Several practical and workable models have been developed to predict the steady-state mean concentration in a lake. The most useful model, admired for its simplicity, is by Vollenweider (1969a, 1975). This is a mass-balance model and treats the lake as a completely mixed system, although there is a modification of the washout coefficient to compensate for reduced flushing during stratified periods. More complex models have been developed by Snodgrass and O'Melia (1975) and Imboden (1974) to predict two forms of P, in two compartments of a lake, and oxygen content. Only Vollenweider's model and modifications will be covered here.

The equation that defines the mass balance of P in a lake, in which losses occur through flushing and sedimentation as a function of P concentration, is:

$$\frac{dP}{dt} = \frac{L}{Z} - (\rho + \sigma)\,P$$

where L is the annual areal loading, ρ and σ are the flushing and sedimentation rates (per year), and P is the concentration of phosphorus.

The corresponding integrated equation that defines P content in a lake at time t is (Vollenweider 1969; Larsen and Mercier 1976a):

$$P_t = \frac{L}{\bar{Z}(\sigma + \rho)}\{1 - \exp[-(\sigma + \rho)t] + P_0 \exp[-(\sigma + \rho)t]\}$$

Thus, knowing the new loading to a lake and the initial P_0 concentration, the expected exponential increase or decrease in P content can be estimated. Although the most difficult variable to define is σ, Dillon and Rigler (1974b) suggest that it can be approximated from the retention coefficient:

$$R = 1 - \frac{\text{annual P output}}{\text{annual P input}} = \frac{\sigma}{\rho + \sigma}$$

Jones and Bachmann (1976) found a mean value for σ of 0.65 for a sample of 143 lakes; σ may also be approximated by $10/\bar{z}$ (Vollenweider 1976).

The sedimentation coefficient, σ, has also been found to be approximately $= \rho^{0.5}$ and R to be $= 1/(1 + \rho^{0.5})$ (Larsen and Mercier 1976a and Uttormark and Hutchins 1978). While these empirical formulations can be useful in estimating a sedimentation term as a function of flushing rate, they may also be difficult to rationalize intuitively. That is, on the one hand the sedimentation coefficient, σ, increases as ρ increases and on the other R decreases as ρ increases. While it is easy to see the latter, because with increased rate of lake water replacement there is simply less time for P to settle, it is not as easy to see the former. However, as ρ increases the *rate* coefficient σ must increase also (although more slowly) because if loss of sediment P is to occur, it must occur faster since the time for sedimentation becomes less. The fact that σ increases more slowly than ρ allows for the fraction retained, R, to actually become less as ρ increases.

Although it is useful to know how fast a lake may respond to a change in P income (Vollenweider 1969a):

$$t_{\frac{1}{2}} = \frac{\ln 2}{(\sigma + \rho)}$$

it is probably more valuable to know the final steady-state concentration:

$$\bar{P} = \frac{L}{\bar{Z}(\sigma + \rho)}$$

or by the alternative formulation (Dillon 1975):

$$\bar{P} = \frac{L(1 - R)}{\bar{Z}\,\rho}$$

There is yet an additional modification (Welch et al. 1973a) that can be made to allow for an internal supply of P, such as sediment release, R_s:

$$\bar{P} = \frac{L + R_s}{\bar{Z}(\sigma + \rho)}$$

It is important to include, in addition to sediment release, the internal supply of P from such sources as fish and macrophyte excretion. Although the accuracy of predicting a new steady-state P level in most cases is probably not better than $\pm$ 50%, enough observations are not actually available at this time to evaluate accuracy. Dillon and Rigler (1974b) showed reasonably good agreement between observed and predicted concentrations in a group of Ontario lakes already at steady state. Uttormark and Hutchins (1978) determined that for 23 lakes from which P had been diverted the Vollenweider model (using $\rho^{0.5}$ as an estimate of σ) accurately predicted the new steady state in 82% of the cases. Table 14 shows results from three lakes, from which P was diverted, with respect to the original Vollenweider (1969a) model.

The results from Lake Washington fit the model quite well, whereas the value in Shagawa still remained 100% too high. Lake Washington is a deep, aerobic lake with little return of P from sediments or littoral; Shagawa, however, is much shallower, with an anaerobic hypolimnion and an extensive littoral area, which apparently account for a sizable internal supply in summer (Larsen and Mercier 1976). Sammamish is also anaerobic, but the sediment source of P is not considered significant because of efficient resedimentation. The predictability of P content in Lake Sammamish results from using annual volume-weighted mean values, which include high, prediversion concentrations during fall turnover. Winter-spring concentrations did not change in contrast to the other two lakes.

Plankton algae and BOD

Algal biomass as an organic matter source is really no different from sewage in its ultimate O_2 demand or BOD. The stochiometry of nutrients versus BOD can be hypothesized from the photosynthetic equations (Stumm 1963). CO_2 is fixed into organic carbon with light and nutrients according to a ratio of 106:16:1, C:N:P, as previously indicated:

$$106 \; CO_2 + 90 \; H_2O + 16 \; NO_3 + 1 \; PO_4 + \text{light energy} \rightarrow$$
$$C_{106}H_{180}O_{45}N_{16}P_1 + 154\tfrac{1}{2} \; O_2$$

The photosynthetic quotient, or PQ, $= 154/106 = 1.45$; experimentally this ratio averages 1.2. In the reverse reaction (respiration) O_2 is subsequently used to convert C_{106} to 106 CO_2, which results in O_2 deficits in lakes and sags in streams. The potential for secondary BOD effect

Table 14. *Comparison of predicted (P_p) versus observed steady-state P (P_0) concentrations during winter before and after a significant diversion of waste water P in three lakes*

	Washington		Shagawa		Sammamish[a]	
	P_p	P_0	P_p	P_0	P_p	P_0
Before diversion	63	63	49	49	33	33
After diversion	22	16	12	25	22	27

[a] Annual means.
Note: Values for σ adjusted to prediversion existing P content.
Source: Edmondson (1972); Larsen and Mercier (1976).

from treated sewage in the form of N and P can theoretically be estimated accordingly: about 75% of the assimilable C that is removed
from sewage in secondary treatment can go back into decomposable
organic matter and BOD in the form of plankton (fixed in photosynthesis) simply from the utilization of N and P that is not effectively
removed by secondary treatment. In turn, the BOD potential of P, if
completely utilized, can be estimated as follows:

$$\frac{154\frac{1}{2} \times 32}{1 \times 31} = \frac{4{,}950}{31} = 160 \text{ mg O}_2 \text{ mg P}^{-1}$$

thus 1 mg P = 160 mg O_2 if all the P remaining from treatment is

Figure 78. Results of a bloom and subsequent decomposition and
nutrient release in a large plastic sphere. (Antia et al. 1963)

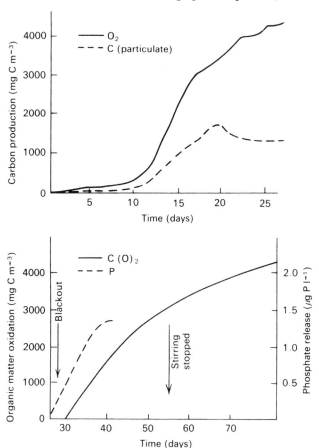

converted to organic C by photosynthesis. In order for that to occur, P must be limiting the plankton growth.

To illustrate the secondary BOD effect, an experiment by Antia et al. (1963) is appropriate (Figure 78). Measurements were made during and following a phytoplankton bloom in a 6 m diameter plastic sphere submerged in the sea. The top graph shows production of carbon by the oxygen method and particulate carbon measurements. The lower graph shows decomposition of produced organic carbon. The oxygen used is expressed as carbon. Note the release of phosphorus after blackout, which caused cessation of photosynthesis and death of the plankton. The greater oxidation of C than production of particulate C is explained as the bacterial breakdown of dissolved organic matter (difference between production as measured by particulate C and O_2). The authors suggest that oxygen consumption is large as a result of the bacterial breakdown of excreted dissolved C and that particulate matter is used more as bacterial substrate.

This process can be observed to cause O_2 problems in overfertilized waters. For example, Baalsrud (1967) showed that the oxygen depletion problem in the Oslo Fjord, Norway, is caused largely by the fertilizing effect of sewage added to inflowing fresh water (which also furnishes needed iron). The dominant organism is *Skeletonema costatum*, a marine diatom that frequently dominates in nearshore waters. This problem can occur because for each gram of phosphorus fixed into algal tissue from treated or untreated sewage at least 50 g of oxidizable carbon is fixed from naturally occurring CO_2 (Figure 79).

Nutrient limitation and plankton growth

The following is offered to substantiate the point already made that the most limiting nutrient in fresh water is P and in marine areas is N. In addition to illustrating the limitation by phosphorus and nitrogen, the following evidence also notes the importance of silica and trace elements in some situations and species. Also of interest is the concept of changing ratios of N:P and changes in which nutrient is limiting over a trophic-state range from oligotrophy to eutrophy.

Silica (SiO_2) limitation of diatom growth

As was pointed out previously, productivity and biomass formation are controlled by the *supply* of the most-limiting nutrient in dissolved form, not the *concentration* at any one time. Of course, there are instances in which the supply is indicated by concentration. Macronutrients that apparently do not recycle rapidly in the epilimnion, such as silica, show limitation when concentration of the dissolved form, which indicates supply, reaches low levels. The blooming of *Asterionella formosa* in Lake Windermere, England, is a clear example (Figure 80).

Laboratory experiments showed that 0.5 mg l^{-1} of SiO_2 is the minimum that will sustain maximum growth of this diatom. The lake concentration is shown as the open horizontal bar. Cessation of growth was directly related to this minimum concentration. This is one of the few instances in which the open-water concentration of one factor has been consistently identified as a limit to growth by a threshold effect

Figure 79. Algal growth potential experiment showing conversion of inorganic nutrients in sewage to oxygen-demanding organic matter. Top graphs show the potential of sewage in Oslo Fjord water at different distances from the sea.

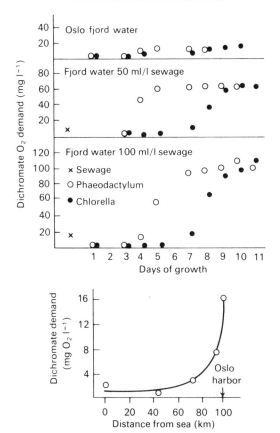

at a relatively high concentration. Although N and P concentrations frequently limit growth rate at very low concentrations, an abrupt effect related to a definite concentration has not been easily demonstrated.

That the 0.5 mg l⁻¹ as a critical level has a fundamental basis was shown by Kilham (1975), who demonstrated that the uptake rate of SiO_2 followed a typical Michaelis-Menton plot and that the levels accounting for 90% of μ_{max} were 0.82 and 0.39 mg l⁻¹, very close to the level determined by Lund (1950) for *Asterionella*.

Nitrogen and phosphorus limitation

Because these nutrients regenerate so rapidly, a concentration of the dissolved form is not indicative of supply. Often, however, the N/P ratio may indicate which nutrient will limit first following continued growth. Figure 81 illustrates the growth of *Selenastrum capricornutum* (green algae) and *Skeletonema costatum* (marine diatom) in culture (Skulberg 1965). The initial N/P ratio in both situations is about 33/1 by atoms. If removed at the ratio of 16 N to 1 P, P will disappear first, and clearly that is what happened in this experiment. In natural waters, however, a parcel of water is not isolated from the environment and nutrient exchanges around that parcel, so concentration in open water does not indicate supply quite as well as in culture. Nevertheless,

Figure 80. *Asterionella* biomass and nitrate and silica content in Lake Windermere in 1945 and 1946. (Lund 1950 with permission of Blackwell Scientific Publications, Ltd.)

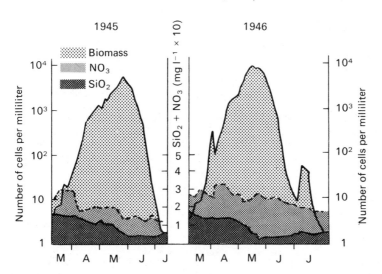

N/P ratios are still useful. Here, the N and P refer to the soluble, usable concentrations.

The effect of the N/P ratio on growth-rate limitation can be illustrated in another way. It can be shown that productivity responds to N and P addition in various combinations in relation to changes in the N/P ratio (Figure 82). This is particularly true if several days are allowed for development of the biomass (Sakamoto 1971).

The limitation in A occurs when N/P goes below 7/1 (by weight) and in B when N/P goes above 7/1. As the curves indicate, there is an interaction, and the ratios do not operate on productivity in an absolute

Figure 81. Growth of two species of algae related to N and P concentrations (soluble) in culture. (Skulberg 1965)

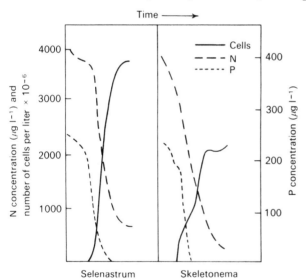

Figure 82. Response of phytoplankton ^{14}C uptake to P and N additions showing the significance of the ratio concept in determining which nutrient is limiting. (Generalized from Sakamoto 1971)

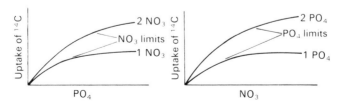

manner. That is, even when the N/P ratio is no longer favorable for growth, as a single nutrient is increased, slight increases in growth may nonetheless occur.

Recall from the discussion of nutrient cycles that, based on the N/P ratio, N usually tends to be limiting in marine waters and P in fresh waters. The N/P ratio varies with trophic state and decreases with increased eutrophication. In oligotrophic lakes N/P ≥ 16/1 (7/1 by weight), so P should limit in the short term. The addition of P may even inhibit growth in the short term if the change is great enough (Hamilton and Preslan 1970). Oligotrophic Lake Tahoe seems to be an exception to this. Goldman and Carter (1965) showed that N was most limiting in short-term carbon-uptake experiments in large plastic columns. Even if P is inhibitory in the short term, allowing a few days for physiological adaptation has resulted in P being more significant than N in oligotrophic lakes (Hendrey 1973).

In a eutrophic lake N/P < 16/1, so N should limit in the short term. In two Wisconsin eutrophic lakes N was shown to be most limiting to a non-N-fixing blue-green alga (Table 15). The case may be different for N fixers, as we will see later.

This is evidence that NO_3 is the most-limiting factor in these eutrophic lakes. Addition of the three elements together (N, P, Fe) resulted in growth about equal to that produced in the synthetic media, which contained a surplus of everything this alga is known to require.

Table 15. *Growth of* Microcystis aeruginosa *during 2 wk in sterilized surface water from two lakes in Wisconsin containing various chemical additives*

Additives	Mendota Lake water (cells mm^{-3})	Percent relative growth	Waubesa Lake water (cells mm^{-3})	Percent relative growth
Without additives	500			
NO_3	5 780	6	1 523	14
PO_4	500	64	7 737	73
Fe	495	6	1 101	10
PO_4 + Fe	400	6	1 139	11
NO_3 + Fe	4 310	4	874	8
NO_3 + PO_4 + Fe	9 030	48	9 675	92
Synthetic nutrient solution	10 235	100	10 575	100
		113	10 235	97

Source: Gerloff and Skoog (1954).

NO$_3$ stimulated the major portion of the growth, as shown by results from those tests in which each of the three elements was used alone. Although NO$_3$ is clearly the most significant factor, there is added stimulation from all three elements combined, with P being secondarily most important.

Even though N can be shown to limit non-N fixers in eutrophic lakes, given sufficient time, P can be seen to be most limiting. That is, the addition of P stimulates N fixation and the biomass increases to match the P addition. In Moses Lake, Washington, NO$_3$ decreases to zero in summer and the addition of NO$_3$ would probably produce the fastest response, but allowing for N fixation, the addition of P alone nearly equals N + P after several days (Welch et al. 1973b) (Figure 83). In contrast, the addition of N alone developed little growth after the existing PO$_4$ was exhausted. Horne and Goldman (1972) found that

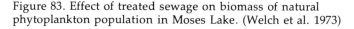

Figure 83. Effect of treated sewage on biomass of natural phytoplankton population in Moses Lake. (Welch et al. 1973)

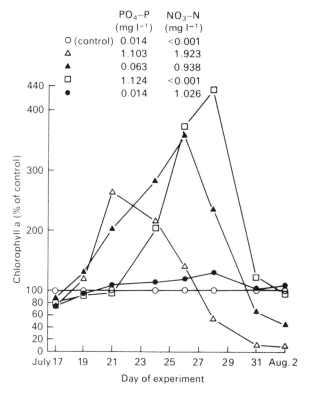

about one-half the N supply to Clear Lake, California, is from N fixation and, although not a fast process, does contribute to the eventual utilization of P. Schindler and Fee (1974) observed that N buildup through fixation, to match a P addition, took from 2 to 3 wk in a plastic-bag experiment.

Table 16 shows that as one observes the N:P ratios either in winter, summer, or annually in lakes there is a tendency for the process of eutrophication to lower the ratio. Thus, lakes move from primarily P limiting through oligotrophy and mesotrophy to N limiting in the mid and later stages of eutrophy.

This tendency was further illustrated by Miller et al. (1974) on 49 lakes in the United States by AGP experiments. Figure 84 shows that as the yield in dry weight of *Selenastrum* increased, the percent of P-limited lakes tested decreased. The yield was taken as indicative of the overall fertility and trophic state.

Two possible explanations for the nutrient limitation response to increased eutrophication have been provided by Thomas (1969) and Hutchinson (1970a). According to Thomas, NO_3 is always observable in oligotrophic lakes, but as PO_4 is added culturally, the "NO_3 reserve" is depleted, probably because the N/P ratio is often low in waste sources. For example, in sewage the ratios are from 2/1 to 4/1. In hypereutrophic lakes in summer, as eutrophication increases, the residual PO_4 content increases until it is relatively high because light becomes limiting from high algal densities. In these situations, NO_3 tends to be most limiting in the short term, but in the long term, PO_4

Table 16. *Annual means of NO_3-N/PO_4-P (by weight) showing decreasing trends as eutrophication proceeds; unless specified, values are annual means*

Kinds of lakes	Location	N/P ratios
Hypereutrophic	Hjälmaren (Sweden)	4/1
	Moses (WA)	1/1 (summer)
Eutrophic	Washington (WA)	10/1 (winter, 1963)
Mesotrophic	Washington	20/1 (winter, 1970)
	Mälaren (Sweden)	25/1
	Sammamish (WA)	27/1
Oligotrophic	Vättern (Sweden)	70/1
	Chester Morse (WA)	63/1

Source: Swedish lakes from National Environmental Protection Board.

is most significant. This was illustrated by the eutrophication and oligotrophication in Lake Washington (Table 16).

Hutchinson suggests that as long as PO_4 is available, N builds up through biological fixation to produce N/P ratios near 16/1. This results in N and P being simultaneously limiting in many short-term experiments, whereas the addition of N or P alone produces little increase in growth. That is the usual condition in mesotrophic and moderately eutrophic lakes. If N alone limits, N fixers must be outcompeted by non-N fixers, for example, Lake Mendota, *Microcystis* (Table 15).

Luxury uptake

At certain times PO_4 can be stored in cells at levels greater than the metabolic demands, particularly when PO_4 is in excess. Thus, when the ambient concentration of the limiting nutrient is low, growth continues, which is probably also true for N, although not as well documented. Because of luxury uptake, growth is probably best related to cell content. One of the first observations of the phenomenon was made by Gerloff and Skoog (1954), who found that *Microcystis* was limited by the following cell concentrations of N and P: P, 0.1–0.14%; N, 4–7%.

Considerable attention has been given to modeling growth rate by separating the response to internal and external levels of nutrient. See Droop (1973) for a discussion of this.

Figure 84. Yield of algal biomass in algal growth potential (AGP) experiments from 49 lakes in the United States showing a decrease in the percent that were P-limited as the yield (and productivity) increased. (Modified from Miller et al. 1974)

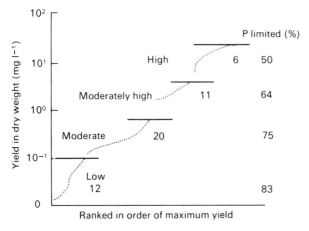

Trace elements and organic growth factors

Many trace elements are required by algae for growth, as cofactors for enzymes, and there is a sound rationale as to how growth rate is affected by trace-element deficiency (Dugdale 1967). Trace-element limitation in natural systems apparently does not occur frequently, however, in view of the fact that full response is nearly always attained with the addition of N and P. Experiments in four lakes, two of which were oligotrophic, in the Lake Washington drainage showed that trace-element additions alone never increased growth, although in some instances an additional enhancement was achieved by a complete medium addition over that of N and P (Hendrey 1973). In view of these results, it would seem that trace elements are usually in adequate natural supply, particularly in eutrophic waters. A notable exception is the productivity response to Mo addition in oligotrophic Castle Lake, California (Goldman 1960a), and some Alaskan lakes (Goldman 1960b). Although many organisms have vitamin demands and may respond to dissolved organic compounds (e.g., glycolic acid, Fogg 1965), these factors have seldom been demonstrated as important limiters to growth under natural conditions.

Chemical complexers

Organic complexers can either remove trace elements from solution and reduce inhibition or make them more soluble and available for growth. The effect of complexers, such as EDTA (ethylene, diamine, tetra acidic acid), has been demonstrated by several workers (Wetzel 1975).

Another factor linked to productivity control is the ratio of monovalent to divalent ions. As that ratio increases, productivity has been observed to increase (Wetzel 1972).

Carbon as a limiting nutrient

Carbon has long been known to be the macronutrient required by plankton in the greatest quantity, although it was not considered seriously as a principal cause for increasing eutrophication until 1969–70. Essentially, the argument centered on the possibility that the organic carbon in waste effluents could cause eutrophication in many carbon-limited situations when the organic C was released by bacterial decomposition as CO_2 (Kuentzel 1969; Kerr et al. 1970). This idea was confronted head on by those who believed very strongly that organic C from sewage or other waste effluents was not a principal cause for eutrophication, notwithstanding the potential limitation by C in some situations.

First of all, it would be useful to mention the points and counter-points made by the two schools of thought in 1970 (Table 17).

To best clarify the problem, it would be most instructive to retrace the steps of this controversy, which originated in 1969 in the article by Kuentzel. He made the following points:

1. Inorganic C in the alkalinity system is insufficient to form a large algal bloom, which frequently occurs in enriched lakes. At pH of 7.5–9, the atmospheric input of CO_2 would amount, at most, to 1 mg l^{-1} after several days of transport.
2. The 56 mg l^{-1} (dry weight) of algae observed in a bloom in Lake Sebasticook, Maine, would require 110 mg l^{-1} CO_2, which

Table 17. *Two schools of thought clash on many points as to whether C or P is the principal key to eutrophication control*

School: carbon is key	School: phosphorus is key
Carbon controls algal growth.	Phosphorus controls algal growth.
Phosphorus is recycled again and again during and after each bloom.	Recycling is inefficient: Some of the phosphorus is lost in bottom sediment.
Phosphorus in sediment is a vast reservoir always available to stimulate growth.	Sediments are sinks for phosphorus, not sources.
Massive blooms can occur even when dissolved phosphorus concentration is low.	Phosphorus concentrations are low during massive blooms because phosphorus is in algal cells, not water.
When large supplies of CO_2 and bicarbonate are present, very small amounts of phosphorus cause growth.	No matter how much CO_2 is present, a certain minimum amount of phosphorus is needed for growth.
CO_2 supplied by the bacterial decomposition of organic matter is the key source of carbon for algal growth.	CO_2 produced by bacteria may be used in algal growth, but the main supply is from dissociation of bicarbonates.
By and large, severe reduction in phosphorus discharges does not result in reduced algal growth.	Reduction in phosphorus discharges materially curtails algal growth.

Source: *The Great Phosphorus Controversy* (1970).

could only come from a supply of organic matter decomposed by bacteria as a ready source of CO_2. About 30 mg l^{-1} of organic carbon would be sufficient for that quantity.

3. The organic C in sewage effluents is furnishing the needed nutrient (C) to form large algal blooms, because such low concentrations of dissolved PO_4-P (< 10 μg l^{-1}) are associated with large blooms. Therefore, it is obvious that sufficient PO_4 exists even at these low concentrations!

Shapiro (1970) responded to this article with the following points:

1. PO_4 is frequently, but not always, the cause of eutrophication.
2. Algae have an absolute requirement of P; 10 μg l^{-1} dissolved at the time of a bloom does not indicate the supply, but rather only the difference between uptake and supply.
3. Sediments are not usually a source for P, but rather a sink.

A report by Kerr et al. (1970) next entered the scene, demonstrating in experimental conditions that C could limit algal production in naturally soft water and that bacteria-originated CO_2 from organic matter serves as an effective source of C in such environments. More specifically, Kerr's findings were briefly as follows:

1. *Anacystis nidulans*, an obligate photoautotroph, was used as the test organism, which attained concentrations of 60 $\times$ 10^6 cells ml^{-1}.
2. Algae that received CO_2 from a dialysis bag with bacteria and organic matter grew to double the mass of that in the flask without such a CO_2 source; 6.5 versus 3.7 $\times$ 10^7 cells ml^{-1}. The same results were produced using CO_2 free air and 5% CO_2 air bubbled into flasks.
3. To determine if such bacteria-originated CO_2 was important in natural systems, and consequently, if the C from organic matter added to natural systems was important in bloom production, various enrichments with "infertile" pond water (< 5 μg l^{-1} P and N) were tested and population increases were recorded after 30 hr of "incubation" (Table 18).

Kerr showed that CO_2 probably limited *growth rate* in the water tested, which was soft. The alkalinity was only 14 mg l^{-1} $CaCO_3$ (DIC = 5 mg l^{-1}), because the addition of organic C to this natural infertile water had about as significant an effect as N and P addition. Heterotrophs and autotrophs were stimulated by both N + P and glucose, but the greatest stimulation occurred with the addition of all three nutrients. Limitation of *growth rate* in these waters was clearly C > P

+ N. However, a question remains. Because glucose addition produced the smallest increase of any combination, would glucose addition produce the greatest long-term effect over one or more years in such infertile water? That is, which nutrient is most scarce (controlling) in the long term?

Schindler (1971b, 1974a) answered the latter and most important question from the standpoint of lake management by a fertilization experiment in a low alkalinity (15 mg l^{-1} CaCO$_3$) lake (Lake 227) in Ontario, Canada. To that system, N and P were added weekly for three summers in an amount to reach 10 μg l^{-1} P. There was no change in total C in the water, and the greatest change in lake algal content did not occur until the third year (Figure 85).

Dissolved P concentration showed very little change in response to fertilization, indicating the supply rate as the significant factor. Even in this low-carbon lake, in which short-term (hours) in situ photosynthesis experiments showed C to limit (Schindler 1971b), a large mass of blue-green algae (*Lyngbya* and *Oscillatoria*) can be produced if the N and P input is maintained long enough.

By following the CO$_2$ diffusion pattern with radium, Schindler (1971b, 1974a) showed that the C supply originated in the atmosphere and that CO$_2$ was made to be limiting in the short term by adding N and P (as Kuentzel hypothesized and Kerr demonstrated). The pattern of CO$_2$ concentration in the fertilized and unfertilized lake over the summer can be illustrated as in Figure 86.

The arrows, which indicate the determined direction of CO$_2$ transfer in the fertilized lake, show that additional amounts of N and P in-

Table 18. *Relative response of* Anacystis nidulans *grown under various conditions*

| Conditions | Natural populations | | Effect of *E. coli* | |
	Bacteria	Algae	CO$_2$ + HCO$_3$ mg l^{-1} increase + 72 hr	CO$_2$ produced (mg) trapped
Pond water	Decrease	No change	3.6	0
+ N and P	Large increase	3 × increase	10.0	0.003
+ Glucose	2 × increase	2½ × increase	9.2	0.005
+ Glucose + N and P	Large increase	6 × increase	12.0	0.030

Source: Kerr, et al. (1970).

creased the supply of CO_2 from the atmosphere, and eventually a large nuisance level of algae resulted. In a lake with higher alkalinity, the problem of a large algal mass may have occurred sooner because C would not have limited growth rate to such an extent in the short term.

In another lake (No. 304) the same gradual result of a large algal biomass from enrichment with P, N, and C occurred. In this instance, however, P input was curtailed and biomass rapidly declined in spite of continually added N and C (Figure 87).

Figure 85. The response of phytoplankton standing crop (as chlorophyll *a*) in the epilimnion of lake 227, fertilized with 0.48 g P and 6.29 g N m^{-2} annually beginning in 1969. Chlorophyll concentration from six other unfertilized lakes never exceeded 10 μg l^{-1}. All lakes are less than 13 m deep. Note that lake 227 had a low standing crop (+) similar to that of the other lakes prior to fertilization. Large inputs of P and N will cause eutrophication problems regardless of how low carbon concentrations are. The necessary carbon is drawn from the atmosphere. (Modified from Schindler and Fee 1974b)

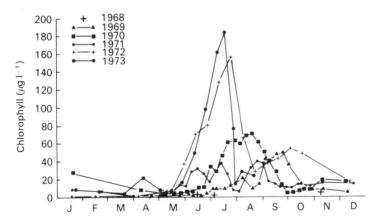

Figure 86. CO_2 level in fertilized and unfertilized lakes compared with the air saturation level of 100%. (Idealized from Schindler 1971b)

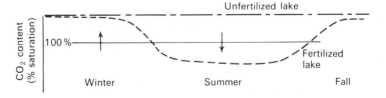

The conclusion with regard to controlling nutrients in eutrophication is that P is usually more significant than N or C in the long term because the P supply relative to the needs for growth is usually less than for N or C. This is because the sources of P are more limited (N and C come from atmosphere), and PO_4 is more readily bound chemically, resulting in sediments usually being more of a sink than a source with respect to P.

However, any nutrient (usually C, N, or P) may limit growth rate in the short term. In this consideration the most-limiting (controlling) nutrient can usually be indicated from the ratio of one concentration to another as indicated previously.

Nutrients and phytoplankton succession

One of the most interesting and challenging subjects in limnology is the succession of species in the phytoplankton. Although this is a problem in its own right, it is also a problem in water-quality management. Succesion occurs seasonally in lakes, as has been discussed previously in relation to tempeature and light. Usually diatoms dominate in the spring, greens in summer, blue-green algae in late summer, and possibly diatoms again in the autumn. However, there

Figure 87. Phytoplankton standing crop (as chlorophyll *a*) in the epilimnion of lake 304 before fertilization (1968, 1969, and 1970); during fertilization with P, N, and C (1971 and 1972); and with the elimination of P and continued fertilization with N and C (1973). Results show the effectiveness of phosphorus control in eutrophication management. (Schindler and Fee 1974b)

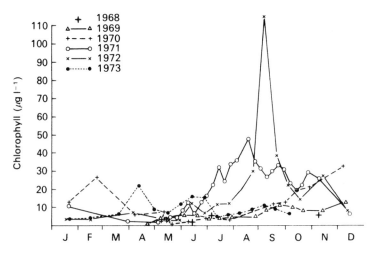

is much variability in this pattern. The variability that is of particular interest here is the tendency for blue-green algae to dominate a greater and greater part of the growing season as eutrophication increases. The objections to the blue-greens are many, but particularly they include: an objectionable appearance because they float and clump together causing scums, objectionable tastes and odors around beaches and drinking water supplies, a reduction in efficiency in predator–prey food chains because the filamentous and colonial nature of blue-green algae preclude effective grazing by filter-feeding zooplankton, and a direct toxicity to mammals and fish when the blooms are very dense.

The challenge to water-quality management is to understand the cause(s) for the domination by blue-green algae when eutrophication increases. If something can be added biologically or chemically to more efficiently utilize the nutrients added to the system, the effects of eutrophication (obnoxious blooms of blue-greens) can be alleviated. This would be particularly valuable in situations where for practical or economic reasons the external nutrient supply cannot be reduced (Shapiro et al. 1975).

The role of light and temperature in the succession of phytoplankton, discussed previously, concluded that those factors, while important, could not begin to explain the phenomenon of blue-green dominance either seasonally or with an increase in eutrophication.

Varying requirements for certain nutrients was early on thought to contribute to succession. The succession from *Asterionella formosa* to *Dinobryon divergens* in the English lakes has been related to the different P requirements of the two organisms being met successively as the P content declines in the spring (Lund 1950). The following ranges of dissolved P in μg l^{-1} for optimum cell growth were suggested by Rodhe (1948):

	Lower limit	Upper limit
Low P		
Dinobryon divergens	< 20	< 20
Urogena americana		
Medium P		
Asterionella formosa	< 20	> 20
High P		
Scenedesmus quadricauda	> 20	> 20

The Chlorococales (e.g., *Chlorella*) apparently require high phosphorus concentration. The situation with blue-green algae (or at least the nuisance species) may be similar to that of either the high, medium, or low nutrient requirement. The bulk of the findings thus far suggests that many of them may do better on low concentrations of dissolved nutrients, yet require high supplies. There are, however, several hypotheses that are all plausible and probably not mutually exclusive.

First, because the blue-green algal requirement for limiting nutrient concentration is less than the predecessor species, their success results from greater efficiency of utilization. This mechanism includes the greater success of N-fixing algae when NO_3 and NH_3 are exhausted. However, probably the most interesting and yet controversial idea has involved the utilization of carbon. The blue-green requirement for CO_2 (free) has been hypothesized to be less than that for green algae, so that when CO_2 is increased, the shift is from blue-green to green algae, and when decreased, the opposite shift occurs. This was first proposed by King (1970) and has later been supported by Shapiro (1973) with results from the experimental setup shown in Figure 88. The experiment was performed in 12 polyethelene cylinders about 1 m in diameter and 1½ m in depth, utilizing natural blue-green algal populations. The various treatments performed are indicated in Figure 88.

Nitrogen and phosphorus were added where indicated in Figure 88 at the rate of 700 μg l^{-1} N and 100 μg l^{-1} P. HCl was added to lower pH to 5.5. The experiment demonstrated that high levels of N and P did not cause a shift from existing blue-green algae, but a shift to green algae did occur when CO_2 was also added. When the free CO_2 concen-

Figure 88. Shapiro's (1973) experiment showing the significance of CO_2 in causing a shift between green and blue-green algal dominance. (Schematic from personal communication)

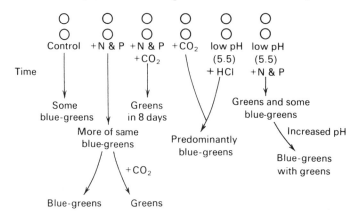

tration was increased by either lowering the pH or adding gaseous CO_2, but without N and P, a shift did not occur to green algae. The shifts to greens and back to blue-greens were also accomplished to some extent when N and P were added and the free CO_2 content was manipulated by decreasing and increasing pH, respectively. The conclusion appeared to be clear. Increased eutrophication with N and P could result in dominance by either green algae (small and more palatable to zooplankton and nonscum forming) or blue-green algae with opposite attributes, depending on the amount of free CO_2. By lowering or raising free CO_2, the outcome in algal dominance could be manipulated and blue-green algae seemed to outcompete greens for the available N and P if CO_2 was low and vice versa if it was high.

Through laboratory and field work with sewage lagoons, King (1970, 1972) had originated the hypothesis that free CO_2 controls blue-green/ green algal dominance. The hypothesis is easily explained by Figure 89. Given a particular water and its associated alkalinity, the algal community would progress from being dominated by *Chlamydamonas* to mixed green algae and finally to mixed blue-greens as the pH increased and the free CO_2 content decreased. Furthermore, the shifts were associated with particular levels of free CO_2. The "critical" level for blue-green dominance was set at 7.5 μm l^{-1} CO_2 (or 0.33 mg l^{-1}).

Figure 89. Boundaries for indicated algal communities in sewage lagoons in relation to alkalinity, pH, and free CO_2. (Modified from King 1970)

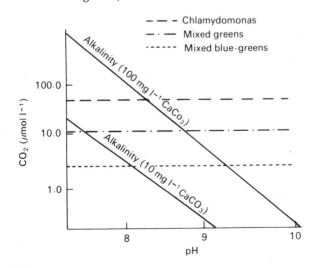

The diagram also shows that for waters of lower alkalinity the critical CO_2 level was reached at a lower pH.

These two aspects can be further clarified by Figures 90 and 91. Figure 90 shows the relationship between the pH, which provides the "critical" CO_2 level, and the dominance of green and blue-green algae and the water's alkalinity. This idea is extrapolated in Figure 91, projecting the amount of P required stochiometrically to remove enough CO_2 to raise the pH to the level providing the critical CO_2 concentration. The amount of P required would vary with the alkalinity of the

Figure 90. The pH at which the critical CO_2 concentration is reached, separating blue-green from green algae as a function of alkalinity. (King 1972)

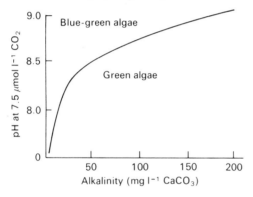

Figure 91. Hypothetical amount of P that must be added to extract enough CO_2 through photosynthesis to, in turn, raise pH to the level providing the critical CO_2 concentration as a function of alkalinity. (King 1972)

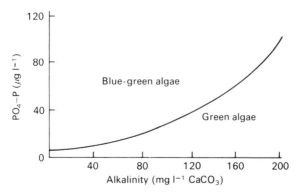

water. Waters of low alkalinity theoretically need much less P enrichment to drive up the pH and, consequently, the free CO_2 level down to the critical level than would high alkalinity waters, because higher alkalinity means more buffering capacity. Although this is a very interesting hypothesis, it needs more verification, particularly with regard to levels involved. The general idea is supported by the experiment by Shapiro (1973) and by Buckley (1971), who made some supporting observations in a plastic-bag enrichment experiment in Moses Lake, Washington. In Buckley's experiment, the growth of blue-green algae was seen to decrease in proportion to the amount of low-nutrient Columbia River water that was added. When NO_3 was replaced, the pH increased and hence the concentration of free CO_2 decreased as the quantity of added N increased. Although added N was expected to favor green algae over the N-fixing blue-greens (mainly *Aphanizomenon*), because greens dominated near a sewage effluent input elsewhere in the lake, this did not occur. Instead, the ratio of blue-greens to greens increased with added N (Table 19). The greens could have been inhibited because CO_2 was held at such low levels, much below King's "critical" level.

The idea that the concentration of free CO_2 is a key factor to the dominance of blue-green or green algae has been challenged by Goldman (1973), who maintained that the free CO_2 concentration is not critical to growth rate. Rather, algal growth is related to C_T. This idea was subsequently challenged by King and Novak (1974). Although the debate was very stimulating, no consensus was really reached. However, it seems that algal growth does respond to free CO_2 because in Shapiro's experiments (1973 and Shapiro et al. 1975), the CO_2 was increased by lowering pH with acid, which in the short term should not have resulted in decreased C_T. Nevertheless, the phenomenon can

Table 19. *The ratio of blue-green cell counts to greens, maximum pH, and minimum CO_2 in 1 m deep plastic-bag experiments lasting $1\frac{1}{2}$ wk in Moses Lake, Washington*

NO_3N mg l^{-1}	0.74	0.5	0.23
Max pH	10.1	9.8	9.6
Min CO_2 mg l^{-1}	0.02	0.07	0.1
B-G:G	1.9:1	1.4:1	0.9:1

Source: Buckley (1971).

be repeated by the manipulation of inorganic C, and whether explained as CO_2 or C_T makes little difference to the final outcome.

The exhaustion of trace elements has also been suspected of favoring blue-green algae while inhibiting diatom growth in at least two cases. Patrick et al. (1969) found that recirculated water through artificial streams allowed for depletion of Mn and a shift from diatoms to blue-greens, whereas once-through water did not allow such a shift. P. Olson (personal communication) found in Fern Lake, Washington, that a "complete medium" fertilization (including all trace elements) resulted in diatom blooms and subsequent zooplankton increase, whereas ammonia super phosphate fertilization resulted in *Anabaena* and relatively little zooplankton.

The uptake rate for PO_4-P (as ^{32}P) has been shown to be lower for blue-greens than for greens as reported by Shapiro (1973). Ahlgren (1977) has found the K_s for growth of *Oscillatoria agardii* to be only about 1 μg l^{-1} P in continuous-flow culture experiments. For maximum growth rate, about 10 μg l^{-1} was required.

According to a second hypothesis, the blue-green algal requirement for limiting nutrient concentration is more than that required by the predecessor species. This has merit because colonies are usually large, even if individual cells are small, and thus the surface/volume ratio is relatively low and should result in a need for high concentrations of nutrient. This plus the fact that they rise near the surface on warm, windless days, resulting in poor circulation, should lead to a need for a high concentration of nutrient. Depleted nutrient concentrations near cell walls are not apt to be readily reestablished under such conditions. Observational evidence indicating that *Oscillatoria* favors high concentrations comes from Edmondson (1969). Figure 92 suggests that *Oscillatoria* showed its greatest rate of increase when the soluble nutrient content was high.

Also note that the standing crop of *Oscillatoria* remained high for 2 or 3 mo even though the nutrient concentration (N + P) was quite low. This indicates that some growth must have been occurring when nutrient content was low.

A third hypothesis is that zooplankton grazing selects for blue-green algae because colonial and filamentous blue-greens, usually with large coverings of mucilage, are too big for filter feeders. This has been shown in field and laboratory experiments by Porter (1972) and by Arnold (1971), who also suspected a toxic inhibition to zooplankton.

Fourth, that excretory products from blue-green algae inhibit other algae and therefore favor the dominance by blue-greens. This has been shown by Vance (1965) and Keating (1977) and was discussed earlier in this chapter in relation to temperature and light.

Fifth, that increased enrichment with N and P depletes silica, which is slow to regenerate, and which, in turn, would limit diatom growth, leaving more nutrients for blue-greens. Schelske and Stoermer (1971) have suggested that southern Lake Michigan is gradually shifting to blue-green algae for that reason.

Sixth, that blue-green algae blooms form near the bottom where nutrient content is higher and then rise to the photic zone with the aid of their gas vacuoles (Fitzgerald 1964; Silvey et al. 1972).

Figure 92. Oscillatoriales growth versus PO_4 and NO_3 concentration in Lake Washington during 1962. The vertical dashed line indicates the nutrient content at the time of maximum biomass. (Edmondson 1969 with permission of the National Academy of Sciences)

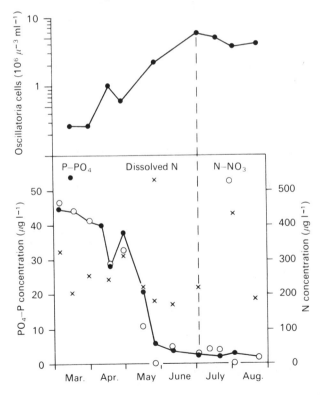

A final theory is that the pseudovacuoles of blue-green algae result in lower sinking rates than other algae. The contrast in sinking rates is most pronounced with diatoms. Knoechel and Kalff (1975) showed that the specific growth rates of a diatom and a blue-green during a typical succession were not different. Rather, they concluded that the reduced mixing as stratification proceeded and the epilimnion deepened was the main factor favoring blue-greens because of sinking rate differences.

Probably several or all of these mechanisms may be operating in any given situation to result in the formation of blue-green blooms and their continued dominance. However, the preponderance of data at this time would seem to favor the hypothesis that blue-green algae, in order to form a large dominant mass, must have a large supply of the most-limiting nutrient, which is usually PO_4. As the supply rate of PO_4 increases, however, the resulting increase in production drives down many soluble nutrient concentrations (e.g., PO_4, CO_2, and trace elements), thus favoring blue-greens with an apparent greater ability to absorb nutrients at low concentration and ultimately to build a large mass from the large regenerated supply rate.

In support of this, large masses of blue-greens have been observed to grow rapidly in the presence of low concentrations of soluble nutrients in Moses Lake, Washington, with growth rate declining only when total N and P decreased (Figure 93). There may be, however,

Figure 93. Maximum growth rate of blue-green algae during the first 4 days of a 2-wk dilution experiment related to initial measured concentrations of NO_3, PO_4, and total P. Total N was estimated from lake concentrations and is 10X the scale values. (Buckley 1971)

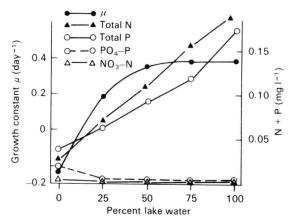

other factors in the dilution water that favor diatoms and discourage blue-greens. These experiments were conducted in plastic bags with the contents exchanged manually at a rate of 10% per day (Buckley 1971; Bush et al. 1972).

With such a mechanism in operation, one can also explain the increase in blue-green algae in Pine Lake in the face of no changes in inorganic PO_4 during the stratified period, whereas a three- to fourfold increase occurs in total P and chl *a* (Figure 94). To be sure, one is not positive which nutrient is most instrumental in allowing blue-greens to dominate. CO_2 concentration is no doubt also very low because of high photosynthetic rates and pH. If King and Shapiro are correct, the addition of acid or CO_2 to eutrophic lakes like Moses and Pine should reduce the pH, raise the CO_2, and favor a switch to green algae. One way of doing this is to completely aerate the water column of a lake in which N and P are rather high (so that vertically transported amounts are not too important) and CO_2 is kept low by high rates of photosynthesis. Aeration will then increase the input of CO_2, lower pH, and greatly raise CO_2 levels. Shapiro et al. (1975) noted a decrease in blue-green dominance in three of six documented cases.

Returning for a moment to the interaction of temperature and nutrient levels on species succession, Hutchinson (1967) hypothesized that blue-green algae occur in eutrophic lakes in summer when soluble nutrient levels are low because these algae are better able to compete with other species for the low nutrient levels at higher, more favorable temperatures. How blue-greens do it under highly enriched conditions during the cooler parts of the season can be hypothesized from work by Goldman and Carpenter (1974) for a blue-green versus a diatom. As shown in Figure 95, the μ_{max} of the blue-green would exceed that of

Figure 94. Total P, ortho P, and chlorophyll *a* in Pine Lake, Washington, during 1971. (Stamnes 1972)

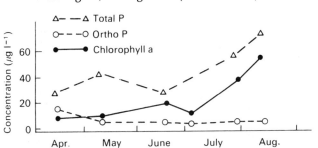

the diatom at high temperature and vice versa at low temperature. If the K_s for the limiting nutrient was less for the blue-green than for the diatom, the two situations of high and low temperature could be represented in Michaelis-Menten fashion (Figure 95). During high temperature, blue-greens are favored over diatoms at low and at high nutrient concentrations. At low temperature, diatoms would be favored only at high nutrient levels; at lower nutrient levels blue-greens would be favored. Although their growth rate would be low, a high supply rate would allow a large biomass buildup of blue-greens.

Lake restoration

The control of eutrophication of standing water, or lake restoration, is a very new but rapidly growing technology. The supply of phosphorus, generally agreed to be the causative nutrient, at least in the long term, should be reduced if improvement in the quality-affected fresh waters is to occur. The controversy and uncertainty arise in the projection of the benefits for any given water body (the degree of improvement) versus the costs to remove the fraction of P loading that can be controlled and by what method. Because experiences have ranged from a high level of success to no improvement at all, there are risks involved when a control program is activated. However, as a variety of experiences are accumulated, the accuracy of estimating benefits as well as costs will improve.

Figure 95. Hypothetical relationship between a diatom favored by high nutrient and low temperature and a blue-green alga favored by high temperature and low nutrient. (Modified from Goldman and Carpenter 1974)

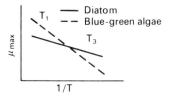

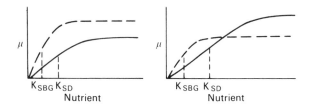

At present, the Vollenweider (1976) model and modifications and the empirical relationships among trophic state indicators provide the most logical approach at estimating benefits from known manipulations. As noted previously, the Vollenweider model, or modifications of it, has been rather dependable in predicting the trophic change in lakes following P diversion (Uttormark and Hutchins 1978).

In order to utilize the P model, however, a reasonably accurate estimate of nutrient loading and outflow is needed. Also needed is an estimate of sedimentation rate. Because that may be too time consuming to measure directly, it can be approximated from a sediment-core analysis or by empirically derived estimates from flushing rate or mean depth as noted earlier. By knowing dry weight, P concentration per unit dry matter, and the accumulation rate of sediment, the flux rate of P to the sediment can be calculated. The rate coefficient for the Vollenweider model can then be obtained knowing the lake concentration.

The rate of sediment accumulation can be determined by associating some historical event with an identifying band in the sediment core. However, in lakes in which nutrient loading has increased recently, the sediment accumulation rate may have changed. Thus, a more accurate measure is by the ^{210}Pb dating method (Schell 1976), which denotes changes in sedimentation rate with time.

Frequently one can work backward from the known lake concentration, flushing rate, etc., and calculate the total loading to the lake. An unknown in the nutrient budget, for example the loading from septic tanks around the lake, in many cases can be approximated by difference.

A variety of techniques for restoring lakes have been thoroughly described and the existing literature on the experiences has been given by Dunst et al. (1974). Some of the more popular techniques with notable, more recent experiences will be reviewed here to suggest the techniques available. The selection of the proper technique depends upon whether the nutrient load is point or nonpoint in character, that is, whether the sizable portion of the nutrient-laden water enters the lake at a single point or at diffuse points or is internal (weeds or sediments). Thus, the techniques will be grouped under point and nonpoint.

Point-source control

The two principal techniques for handling point-source P loading are diversion away from the water body and advanced waste-water treatment for the removal of P, usually by precipitation with alum. Diversion has usually involved the construction of a large interceptor sewer through which the waste water is pumped to either a large treatment plant or a larger body of water with more assimilative capacity. In several instances, problems developed later in the new receiving water. Thus, the diversion technique is appropriate only where the new receiving water is suitable, that is, where nutrients are not the limiting factor.

The main difficulty with advanced waste-water treatment is disposal of the chemical sludge resulting from the precipitation of PO_4. Although the residual concentration of P can be experimentally lowered to 50 μg l^{-1} by alum precipitation, practical plant operation actually leaves about 500 μg l^{-1}. Whereas that is still a 90% or better reduction from secondary treatment effluent, it still may represent too high a loading for a lake, particularly if there are no other lower nutrient sources for dilution.

Diversion. Although diversion has been practiced on several lakes throughout the world, the monitoring record from most is rather poor. Rock (1974) has summarized some qualitative results from 16 lakes (Table 20). In general, definite improvements were noticed in only 5. Nearly all the lakes that showed poor or no improvement were shallow, which no doubt emphasizes the increasing importance of the sediment-originated internal P source as depth decreases. Ryding and Forsberg (1976) have also suggested that reduction in nutrient input may not improve heavily loaded shallow lakes.

The diversion of waste water from Lake Washington by the Municipality of Metropolitan Seattle is the most celebrated case of lake recovery in the world. The Lake Washington recovery, which has been described by Edmondson (1970, 1972), involved the diversion of two-thirds of the P loading from the lake during the period 1963-7. The most significant effects of that recovery are shown in Figure 96 with the comparison of relative values of total P, PO_4, NO_3, and chlorophyll before and after diversion with values also included from 1933. PO_4 and chl a have maintained a rather steady-state level of about 25% of the 1963 values since 1970.

Lake Sammamish, just east of Lake Washington, has shown no response, in terms of winter P or summer chl a, through 1975 to a diversion of about one-third of the P supply in 1968 (Figure 97). The rapid response in Lake Washington is attributed to the aerobic hypo-

Table 20. *Representative lakes in which sewage has been diverted*

Lake	Mean depth (m)	Results of diversion
d'Annecy, France	41.5	Improved following diversion (Laurent et al. 1970)
Fairmont lakes (5 lakes), Minn.	2–3	Slight improvement with diversion but dredging provided remarkable improvement (Ketelle and Uttormark 1971)
Jordan, Mich.	16 (max)	Secondary effluent removed in 1971 (Ketelle and Uttormark 1971)
Lansing, Mich.	3	Remains eutrophic, dredging planned (Ketelle and Uttormark 1971)
Lyngby-Sø, Denmark	2.8 (max)	Good recovery, but increased macrophytes (Edmondson 1969)
Madison lakes, Wisc.	9–25	Only slight improvement (Fitzgerald 1964)
Norrviken, Sweden	5.4	Slight improvement, but P concentration still remains high (Ahlgren 1978)
Rotsee, Switzerland	.2	Diversion in 1933, but no improvement (Edmondson 1969)
Sammamish, Wash.	17.7	No change in winter P, slight reduction in blue-green algae (Welch, 1977)
Schliersee, Germany	40 (max)	Partial diversion in 1964, reduced oxygen depletion by 1967 (Liebmann 1970)
Snake, Wisc.	7 (max)	No improvement (Born et al. 1973a)
Stone, Mich.	20 (max)	Improved only after addition of fly ash (Tenney and Echelberger 1970)
Tegernsee, Germany	70 (max)	Improved oxygen conditions, *Oscillatoria* still present, reduced numbers (Liebmann 1970)
Trummen, Sweden	1.1	No recovery until dredging (Björk et al. 1972)
Washington, Wash.	32	Excellent recovery (Edmondson 1970)
Zurich, Switzerland	50	Still eutrophic (Edmondson 1969)

Source: Rock (1974).

limnion and great depth, which allow very little return of sedimented P, and the rather high water exchange rate (ρ = 0.3 per year). Lake Sammamish also has a high water exchange rate (ρ = 0.55 per year), but the hypolimnion becomes anaerobic prior to turnover, allowing large release rates of P and Fe from the sediment. Following turnover and the establishment of aerobic conditions, precipitation of PO_4 takes place in association with Fe, which means that the sediments are probably controlling the P concentration in lake water and that diversion of one-third of the P supply has not affected that control. Thus, the P released from sediments reenters the sediments again shortly after turnover and the vertical mixing of the released P. Algae growing in the spring and summer actually see little of the anaerobically released P because it is so quickly and efficiently resedimented (Emery et al. 1973; Rock 1974; Birch 1976; Welch 1977). Although the P concentration at turnover, as well as the annual mean value, has decreased since diversion (Table 14), the loading change has probably not been large enough to overcome the control of lake P content by the iron cycle. Effects of the relatively small fraction diverted may have been masked

Figure 96. Relative values in surface water (upper 10 m) of phosphate-phosphorus and nitrate-nitrogen (January to March), total P (whole year), and summer (July and August) chlorophyll *a* in Lake Washington. The 1963 mean values, plotted as 100%, were, in micrograms per liter: Total P, 65.7; phosphate-P, 55.3; nitrate-N, 425; and chlorophyll *a*, 34.8.*1933 data. (Edmondson 1978)

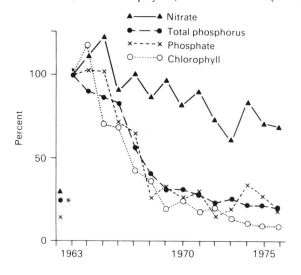

by normal year-to-year variations in winter P and summer chl *a*. Stewart (1976) has suggested that normal year-to-year variations in oxygen and Secchi disk in the Madison lakes are too great to detect any response to nutrient-loading changes. There did, however, seem to be a significant reduction in the fraction of blue-green algae in Lake Sammamish following diversion.

Lake Norrviken has responded predictably to a large reduction in P loading in terms of P concentration (Ahlgren 1978). From 1969 to 1975 total P decreased from 470 to 150 μg l^{-1}, but as would be expected only a slight change has been seen in the trophic state of the lake because the level of P is still so high.

Figure 97. Mean concentrations in Lake Sammamish (usually upper 8 m) of growing season chlorophyll *a* (March–August) and winter (December–February) total phosphorus and nitrate-nitrogen relative to prediversion 1965 levels. The 1965 levels were: chl*a*, 6.5 μg l^{-1} (actually a mean of 1964 and 1965 data); total P, 31 μg l^{-1}; NO$_3$-N, 390 μg l^{-1}. The percent of blue-green algae of the total phytoplankton volume was compared against the prediversion mean for June–October in 1965 and July–October in 1964 (67.5%). (Welch 1977)

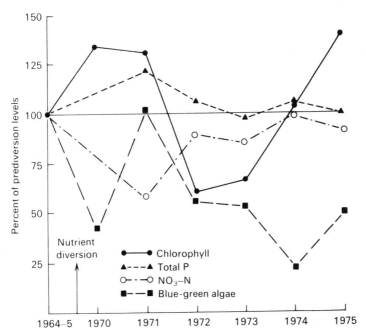

In 1958 sewage effluent was diverted from Lake Waubesa in the Madison lakes chain into the Yahara River downstream from the lake. The nutrient supply to these lakes before (1940s) that diversion is shown in Table 21.

Sewage contributed 75% of the N and 88% of the P to Lake Waubesa. The effect of the diversion from Waubesa on winter soluble P was a reduction from concentrations of over 500 μg l^{-1} to about 50 μg l^{-1}. Although little effect was noticed on the algal biomass in Lake Waubesa, dominance changed from 99% *Microcystis* to a more mixed assemblage (Fitzgerald 1964). This result seems to be similar to that in Lake Norrviken; P concentration still remains at a eutrophic level.

Advanced waste-water treatment. The best test case for advanced wastewater treatment of a point-source P input is at Shagawa Lake, Minnesota (Larsen et al. 1975; Larsen and Mercier 1976b). In 1973 70% of the P input was removed by installing P removal by alum precipitation in the sewage treatment at the town of Ely. Two years after treatment began the winter concentration of total P dropped by 50%, from about 50 to 25 μg l^{-1}, and summer chl *a* from about 65 to 22 μg l^{-1}. Because the lake has a high flushing rate ($\rho \simeq 1.5$ per year), the response has been rather quick. There remains, however, a very large internal input of P in the summer, which is still maintaining the lake as eutrophic. This internal source is apparently the anaerobic sediment and/or the littoral-rooted macrophytes.

Other than the Lake Erie basin, there has been little progress toward the installation of PO$_4$ removal from sewage effluent in the United States. In many cases the approximate doubling of cost for PO$_4$ removal

Table 21. *Nutrient loading to the Madison lakes (in grams per square meter per year)*

	N	P
Mendota	2.2	0.07
Monona	8.8	0.90
Waubesa	47.0	7.0
Kegonsa	6.8	3.9
Koshkonong	9.8	4.3

Source: Lawton (1961).

over that of biological treatment alone has been the cause, particularly because a guarantee of a prompt and significant recovery in the affected lake is not always possible. In contrast, by 1974 the fraction of sewage treatment plants with PO_4 removal in Sweden had increased to more than 50% and 20 such affected lakes are under investigation there (Ryding and Forsberg 1976). The Shagawa Lake case should go a long way toward instilling confidence in the improvement potential by removing only PO_4 from sewage effluents, provided the loading reduction is sufficient.

Non-point-source control

Dilution. The principle here involves the removal of nutrients by adding low-nutrient water and washing the nutrients through the system before algae can utilize them. Because cell washout significantly reduces algal biomass only at a relatively high rate of water exchange, reduction in nutrient concentration and the resulting high nutrient washout are the significant factors that improve water quality. Cell washout is usually not too important because of the high growth rate. Green Lake in Seattle is the best example of restoration by dilution. City water has been added to the naturally eutrophic lake since 1965. The procedure was proposed by Sylvester after an initial study of the lake (Sylvester and Anderson 1964). The addition, which raised the flushing rate to about three times per year, quickly changed the lake to a mesotrophic state. Secchi-disk depth changed from 1 to 4 m, chl *a* from about 50 to 10 μg l^{-1}, and total P from about 60 to 20 μg l^{-1} in 3 yr (Oglesby 1969).

Another example is Moses Lake in Washington, a highly eutrophic lake because of high nutrient loading from irrigation return water, which has ready access to low-nutrient Columbia River water. The first documented test of dilution occurred during the spring and summer of 1977 when Moses Lake received water in three different intervals. Whereas the P concentration in the lake is normally between 150 and 200 μg l^{-1} and chl *a* around 100 μg l^{-1}, the dilution-water addition quickly brought the P concentration to near 30 μg l^{-1} and chl *a* to less than 10 μg l^{-1}. Although the cessation of dilution-water addition would allow the P and chl *a* content to gradually return to a eutrophic state, the levels were held near 50 and 20 μg l^{-1}, respectively, for much of the summer. Quality improved even more following additions in the spring of 1978 and persisted more in spite of no additional summer inputs. The benefits in terms of reduced quantities of blue-green algae have appeared to occur from the inhospitable "new" water in spite of P content still remaining high.

The short-term effect on nonconservative P in such dilution experiments was predicted quite accurately with the following equation:

$$C_t = C_i + (C_0 - C_i) e^{-Kt}$$

where C_i is the inflow concentration, C_0 is the initial lake concentration, C_t is concentration at time t, and K is the flushing rate. Obviously, from the equation, the greater the difference between C_0 and C_i, the better the chance for dilution to restore the lake. The effect of "new" water alone even if P is not greatly reduced may in some instances discourage blue-greens as has been noted in Moses Lake. This effect appears to be consistent with the hypothesis of blue-green dominance by inhibitory products (Keating 1977).

Dredging. In some instances, shallow lakes with rich sediment that show a decreasing gradient in nutrient concentration with increasing depth of sediment may respond to dredging. The best-known example of this technique is Lake Trummen at Växjö, Sweden. As noted earlier (Björk 1972, 1974), diversion of sewage effluent after 60 yr of input had not improved the lake 10 yr later. Removal of 1 m of rich sediment from the bottom of the shallow (2 m) lake quickly resulted in a dramatic reduction in nutrient and algal content (Figure 98).

The principal drawback to dredging is the expense. Lake Trummen is a 100 ha lake and the cost in 1970 was 0.5×10^6. Compared with costs today that was very low. Another limitation to dredging is the necessity for the nutrient content of the sediment to decrease with depth so that relatively nutrient-poor sediment can be reached.

P inactivation. This has become one of the most popular techniques. The complexing substance is usually alum, although such materials as fly ash and zirconium have also been used (see Dunst et al. 1974). One application may last a couple years, and, as might be expected, results are much better in stratified lakes than in shallow polymictic lakes. Probably the best-known examples are Snake Lake, Wisconsin (Born et al. 1973a), Liberty Lake, Washington (Funk, personal communication), and Dollar and Twin lakes in Ohio (Cooke et al. 1978). Experimental work on Clines Pond, Oregon (Peterson et al. 1976), has also shown the effectiveness of removing PO_4 from lake water with $ZrCl_4$.

Smothering effects on benthos and toxicity to fish have been suggested as possible disadvantages of alum treatment in lakes. However, no adverse effect of the alum flock, which settled to the bottom of Liberty Lake following treatment, was noted by Funk (personal communication). Although the flock may be several centimeters thick following treatment, it disappears into the sediment in a few months through bioturbation. Further, toxicity to fish would seem to be a rather remote possibility, largely because the residence time of the material in the water column is short. The alum dosage selected for a lake should produce a residual not to exceed 0.05 mg l^{-1} dissolved Al, which is the amount tolerated by trout over the short term. Because the dosage varies with initial alkalinity, jar tests can be conducted to determine the maximum alum dose. Kennedy and Cooke (1974) found that about 15 mg l^{-1} Al was needed for an alkalinity of near 100 mg l^{-1} $CaCO_3$.

Cycling waste-water nutrients into higher trophic levels. Rather than getting rid of nutrients, another approach is to utilize them biologically. One project has succeeded in growing oysters in ponds with waste-water-produced algae, which results in a highly polished effluent in

Figure 98. Total P and algal biomass (fresh weight) determined at 0.2 m in Lake Trummen before and after dredging. (Bengtsson et al. 1975; Cronberg et al. 1975)

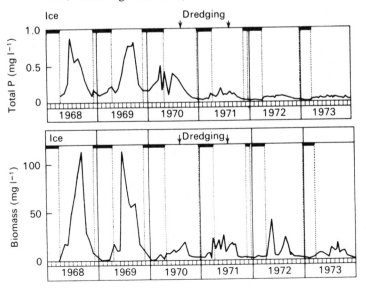

terms of N (Ryther et al. 1972). Inorganic N is 97% removed and P 35% removed from the final effluent by macrophytes. Because N is considered most limiting in seawater, tests with this effluent versus secondary effluent on marine plankton showed a much reduced stimulating effect.

Shapiro et al. (1975) have suggested ways to manipulate the predator-prey food chain to reduce algal biomass. Some manipulation would probably be necessary to change the algal species to edible types (non-blue-green); the presence of *Daphnia*, for example, has a very significant effect on algal biomass. *Daphnia*, a choice food for perch and other secondary consumers, would be better able to control algae if predators to control secondary consumers were introduced or protected. This will be discussed further in the section on zooplankton.

Hypolimnetic aeration. Anaerobic hypolimnia of lakes can be aerated to reduce the P supply from sediments by a unit called a "limno" (Figure 99), which allows the circulation of hypolimnetic water oxygenated with compressed air. However, the air is exhausted back to the surface and the lake is not destratified. The objective is to oxidize iron to the ferric state and complex PO_4, thus reducing the internal supply of P. The technique is now or has been in operation in nine lakes in Europe (Björk 1974). Data are available from two lakes, Järla Lake in Sweden (Bengtsson and Gelin 1975) and Grebiner Lake in Germany (Ohle 1975). Although total P was reduced in Järla from a level around 750 μg l^{-1} before aeration to around 500 μg l^{-1} during aeration, once aeration ceased, the P content returned to the normal level.

Figure 99. Diagrammatic view of a "limno" hypolimnetic aerator (Atlas Copco Airpower N.B., Wilrijk, Belgium). Arrows indicate air direction.

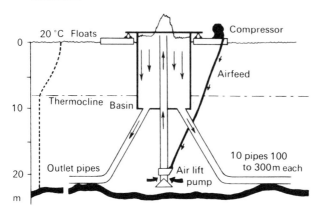

In addition to aeration of the hypolimnion in Grebiner Lake, alum and benthonite were also distributed with the aerated water. Although phytoplankton biomass and productivity were reduced in the water column by 50%, it is not clear how much reduction was due to aeration alone (Björk 1974).

Hypolimnetic withdrawal. This technique, which discharges the nutrient-rich bottom waters from the lake to effect an increase in the loss rate of nutrients, is appropriate only in the case of stratified lakes that have anaerobic hypolimnia in which PO_4 has reached high concentrations. The procedure requires the placing of a pipe from near the deepest point in the lake to the outlet and siphoning off the hypolimnetic water at the rate of the inflow. The pipe is called an Olszewski tube after the original designer. Although some weakening in the stratification will occur, the lake will probably not destratify, the hypolimnion will become more oxygenated, increasing the area for optimum fish growth, and the availability of P to growing algae will decrease upon destratification and for the following spring.

A notable mechanical advantage of this technique is the dependability of the siphon. Because it operates by gravity, there should be no problem from power failure or mechanical malfunction. A pump can be installed as an auxiliary mechanism to increase the discharge if desirable. Such an auxiliary pump was included in a hypolimnetic withdrawal project proposed on Lake Ballinger in Seattle.

A possibly important disadvantage, however, is the effect of high PO_4 and low O_2 levels in the discharge stream. If that stream contains an important fishery or if it is otherwise used for recreation, special precautions must be taken. By comparing the O_2 deficit rate in the lake with the inflow O_2 loading, the extent to which O_2 will be reduced can be determined. If O_2 remains a problem in the outflow, special aeration equipment will be needed.

Whether high PO_4 concentrations will cause a problem is difficult to determine. One approach is to proceed with the discharge and monitor algal activity in the discharge stream. If PO_4 is not limiting growth, a problem should not develop. If the O_2 level is maintained above 3 mg l^{-1} in the hypolimnion, the discharge PO_4 content should approach that of the inflow.

There have been nine Olszewski tubes reported in operation in Europe (Björk 1974). The most notable results, however, are those by Gächter (1976), who has shown extensive improvement in a 51 ha, 3.9

m (mean) deep eutrophic lake in Switzerland. Phosphorus content in the summer decreased drastically, Secchi disk increased significantly, total N decreased, and O_2 increased. *Oscillatoria rubescens* decreased from 152 to 41 g m^{-2}. A mass balance showed that during June and July before withdrawal the sediment supply of P was as much as 275 times the external P loading, but after withdrawal it was reduced to only 4 times the external loading. There was a steady decline in the sediment yield of P and improvement of lake quality for six years after installing the siphon.

Although many of the in-lake treatment techniques have been successful, it is generally agreed that before in-lake treatments are initiated the external loading should be reduced. Where sewage effluent can be diverted or PO_4 removal facilities constructed, this can be readily accomplished. However, many lakes in urban and rural settings alike are receiving increased nutrient loading in runoff from residential and commercial developments and increased fertilization of agricultural and forest lands. Such increased nutrient loading is from nonpoint sources. The only procedure that results in reduced loadings from those sources is improvement of the watershed, thus lessening the runoff potential. Improvements to the watershed are more difficult to initiate and are not likely to show dramatic results in the receiving lake. They are, however, the proper approach to sound lake management in the long term.

See Chapter 8 for study questions on plankton.

8

ZOOPLANKTON

Zooplankton are the primary consumers of phytoplankton in lakes, the ocean, and deep, slow-moving rivers. They also consume bacteria and detritus. The availability of these different foods and the relative abundance and production of different size fractions of food apparently are important factors in the changing pattern of zooplankton production observed in the eutrophication process.

Considerable work has been done on the effect of heat on zooplankton survival. Zooplankton, particularly *Daphnia*, have been used as bioassay organisms to test the toxicity of various compounds. Because zooplankton populations are so dynamic they have not been used extensively as indices of water-quality change. Nevertheless, zooplankton are an important functional and structural component of aquatic ecosystems and their role should be considered in regard to waste effects.

Life cycle

The growth rate of zooplankton is relatively rapid because these organisms are equipped with special reproductive techniques (Figure 100). Rotifers, or the wheeled organisms, are parthenogenic; that is, adults produce eggs that have a full complement of chromosomes (diploid) and therefore require no sexual phase in their reproduction. The crustacean zooplankton have more complicated life cycles and include the two orders Cladocera and Copepoda. Cladocerans (such as *Daphnia*) are also parthenogenic, with males arising from diploid eggs only in instances of crowding, followed by sexual reproduction (haploid sperm and haploid egg) giving rise to resting eggs. Copepods (such as *Diap-*

tomus) have no parthenogenic phase but can also repopulate rapidly because the female can store sperm for several fertilizations from one copulation. Reproduction is totally sexual, and several identifiable stages occur in their development.

Filtering and grazing

Consumption of phytoplankton by filter-feeding zooplankton can significantly control the biomass of phytoplankton, assuming the algal cell sizes are not too large. If one assumes a steady-state condition in which the grazing or removal rate equals the growth rate of the phytoplankton and then increases the grazing rate, the potential impact of grazing on the phytoplankton stock can be illustrated (Sverdrup et al. 1942). From that steady-state situation assume that the grazing is doubled: Then phytoplankton stock will decrease from 10^6 cells to 27 $\times$ 10^3 cells in 5 days. If grazing is tripled, then phytoplankton stock will decrease from 10^6 to < 1 in 5 days.

Measurement of grazing involves a determination of the filtering rate in l per animal per hour and is estimated from short-term experiments in which the amount of algae is determined either before and after or

Figure 100. Reproduction cycle in three groups of zooplankton: rotifers, cladocerans, and copepods.

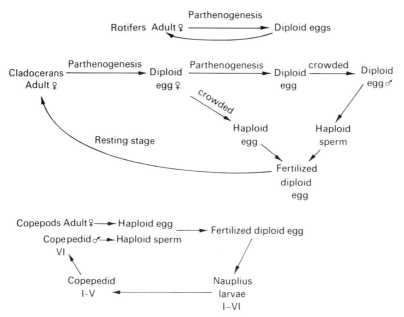

with and without grazing. The removal can also be determined by measuring the uptake of a radioisotope-tagged algal crop by zooplankton over a short time interval (minutes).

The formula for calculating the filtering rate is given by (Rigler 1971):

$$(1 \text{ animal}^{-1}) \left(\frac{\ln C_0 - \ln C_t}{\text{hr}} \right)$$

where C_0 is the initial or control concentration and C_t is the final or test concentration. The grazing or feeding rate is in turn estimated by multiplying the filtering rate by the biomass of algae:

$$1 \text{ animal}^{-1} \text{ hr}^{-1} \times \text{mg C l}^{-1} \text{ (algae)} = \text{mg C (algae) animal}^{-1} \text{ hr}^{-1}$$

The product of feeding rate and animal concentration in the lake gives units that are comparable to primary productivity. This value, in mg C m^{-2} per day, can be directly compared with primary productivity as an indication of the impact of grazing on phytoplankton. In many environments the consumption rate has equaled the phytoplankton productivity rate (Wright 1958; Green and Hargrave 1966). In other environments, usually the more eutrophic waters, phytoplankton production exceeds zooplankton consumption. This occurs because of the growth of large algae that are relatively unutilized by filter-feeding zooplankton, for example, large diatoms like *Asterionella* and *Melosira* and colonial and filamentous blue-green algae. In other situations, tremendous zooplankton production is unrelated to phytoplankton production because of large influxes of detrital allochthonous matter.

Within edible-size limits, zooplankton will eat particles of varying size, and the intake will be related to the volume of food ingested (Table 22).

Table 22. *Maximum feeding rates of* Daphnia *using* ^{32}P *labeled organisms*

$\times 10^6$ cells animal^{-1} hr^{-1}	Particle size in μ^3	Volume consumed in mm^3 animal^{-1} hr^{-1}
5.6 *E. coli* (bacteria)	0.9	0.005
0.5 *Chorella* (green algae)	34.0	0.017
0.25 Yeast	66.0	0.016
0.0028 Protozoan	1.8×10^4	0.051

Source: Modified slightly from McMahon and Rigler (1965).

Within the range of edible-size fractions, the feeding rate is usually considered a function of food density and temperature. With respect to food density, a relation similar to that which represents nutrient uptake by phytoplankton is often considered operative (Figure 101).

This relationship shows, of course, that at a high food density the removal rate by zooplankton is saturated and that further increases in biomass would not be utilized. Coupled with the low edibility of some eutrophic species (blue-greens), these factors suggest a mechanism to allow accumulation of algal biomass in eutrophic lakes.

Measurement

Zooplankton is usually collected with nets calibrated to filter a known volume of water either vertically or horizontally. Biomass is determined usually by counts of the various representatives in their particular life stages. With average weights of the various life-stage representatives, the total mass can be calculated. Production can be estimated by analyzing the biomass and the observed rate of increase. For more details on methods of determining zooplankton production see Edmondson (1974).

Zooplankton and eutrophication

In some instances zooplankton composition has been shown to change in response to eutrophication. For example, *Bosmina corogoni* has been replaced in some situations by *B. longirostris* (Hasler 1947). *Limnocalanus* has nearly been eliminated in Lake Erie (Gannon and Beeton 1971). Much of this change is thought to be related more to changes in fish feeding and to selection of large-sized zooplankton than directly to eutrophication per se. Gannon and Beeton suggest that with *Limnocalanus*, decreased hypolimnetic oxygen forced the organism from preferred lower-temperature water into areas where it was

Figure 101. Hypothetical relationship between grazing or feeding rate of zooplankton and the concentration of food.

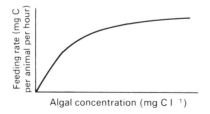

more susceptible to the grazing by populations of yellow perch, which became more dense after eutrophication.

Other experiments have shown large zooplankton to be selectively removed over smaller zooplankton (Brooks 1969). Edmondson (personal communication) has observed *Daphnia* to return to Lake Washington following the recovery from eutrophication. However, their return did not occur until five years after P and chl *a* had equilibrated. The advent of *Daphnia* resulted in marked improvement of water clarity, apparently because of the increased grazing activity. Thus, zooplankton composition may change with enrichment, but the changes may be prompted more by predator–prey relationships than by nutrient or phytoplankton status.

An experiment in a large number of ponds by Hall et al. (1970) showed that regardless of the enrichment (N + P) level the presence of fish (bluegill sunfish) determined the zooplankton composition. That is, the large *Ceriodaphnia* were removed by fish. The zooplankton biomass was 53% *Ceriodaphnia* without fish and only 3% with fish, regardless of whether the nutrient content was low, medium, or high.

However, the effect of fish on the amount of biomass was felt at low and moderate enrichment levels; at high enrichment levels fish apparently had no effect on the biomass of zooplankton (Table 23).

Thus, it would appear that if enrichment is great enough, production and biomass of zooplankton will increase, with the species composition of zooplankton being affected only by fish predation and selection. At moderate levels of enrichment, fish affect both the biomass and the species composition of zooplankton. With increased enrichment of a mesotrophic lake, for example, fish could be expected to remove more of the increased zooplankton production than in an already highly enriched lake.

The effect of fish selectivity on zooplankton can be carried through

Table 23. *Relative biomass level as a function of enrichment and fish*

	Low nutrient	Moderate nutrient	High nutrient
Wo/fish	1.0	2.0	3.0
W/fish	0.5	1.0	3.0

Source: Idealized from Hall, et al. (1970).

to the algae as shown by Shapiro et al. (1975). In experiments with natural plankton populations, Shapiro showed that algal biomass could ultimately (after 50 days) be reduced to low levels by grazing *Daphnia* whether the water was enriched or not as long as perch were absent (Figure 102). These experimental results and the observations on Lake Washington suggest that a considerable potential exists for lake quality enhancement by the manipulation of predators.

With regard to eutrophication, Hillbricht-Ilkowska (1972) suggested a comprehensive hypothesis to account for the relative condition of the plankton expected in the differently enriched environments (Table 24).

Gliwicz (1975), Gliwicz and Hillbricht-Ilkowska (1973), and Pederson et al. (1976) have also shown evidence that favors this scheme. Gliwicz has shown that the filtering rate increases from ultraoligotrophy through mesotrophy but decreases in eutrophic lakes. The zooplankton crop increased through mesotrophy but failed to increase further in eutrophic lakes, thus the efficiency of phytoplankton removal was poor, with eutrophy resulting from an increasing amount of net phytoplankton and bacteria/detritus. Decreasing efficiencies with increased enrichment are also shown for experimental ponds and a series of three lakes (Table 25).

The main point of this hypothesis is that the increased production from eutrophication is accompanied by a decreased efficiency of energy and nutrient utilization in the plankton food web. Although still not adequately proven, it seems reasonable to expect that the poorer utilization of the algal crop as eutrophication proceeds means that the returns to herbivores, and ultimately to the prized predator fishes, will

Figure 102. The control of algal biomass by *Daphnia* with and without presence of planktivorous fish. (Shapiro et al. 1975)

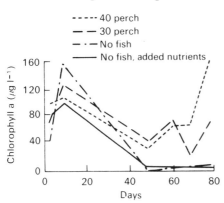

Table 24. *Expected character of phytoplankton-zooplankton relationships with respect to trophic state*

Variable	Oligotrophic-mesotrophic, moderately fertilized, pelagic-marine	Eutrophic, heavily fertilized ponds, neritic-marine
Ratio of zooplantkon consumption/ primary production	≃100%	≃30%
Efficiency of energy conversion	≃20%	≃10%
Size of zooplankton	large	small and do not control phytoplankton
Size of phytoplankton	nanoplankton < 50μ	nanoplankton not used, bacteria and detritus consumers dominate

Source: Hillbricht-Ilkowska (1972).

Table 25. *The ratio of zooplankton to phytoplankton productivity as a measure of food-web efficiency in experimental ponds and three lakes*

Nutrient status	Zooplankton/ phytoplankton production
Ponds	
High	0.07–0.05
Medium	0.41–0.08
Low	0.56–0.20
Lakes (2 years)	
Sammamish – mesotrophic	0.04–0.04
Chester Morse – oligotrophic	0.09–0.08
Findley – oligotrophic	0.18–0.08

Source: Hall et al. (1970); Pederson et al. (1976).

diminish to the point that little of the increased primary production is transferred to fish between the eutrophic and the hypereutrophic state. The goal for biological control of eutrophication is, of course, to understand phytoplankton species succession well enough to effect an adequate change back to edible forms so that the zooplankton can convert more of the nutrient into fish flesh and also provide predator populations adequate to control the prey. Oligotrophic and mesotrophic systems obviously have better balanced predator–prey populations, and a symptom of eutrophic systems is natural imbalance. Thus, the task of biological control of eutrophication is a difficult one.

Zooplankton and temperature

The lethal temperature for zooplankton is usually between 28 and 35 °C (Welch and Wojtalik 1968). Acclimation seems to have little influence on lethal temperature. Heinle (1969a) has shown that for *Eurytemora affinis*, an estuarine copepod, the lethal temperature is between 28 and 32 °C regardless of whether acclimation was 15, 20, or 25 °C. The acclimation time in that experiment was 2 mo or 1 generation time. The rather absolute nature of the lethal temperature for zooplankton will be illustrated again in the discussion of the passage of invertebrates through power-plant cooling systems.

The effect of mild increases in temperature from the discharge of heated water could benefit the productivity of lake systems. For most lakes in temperate areas there would seem to be a rather large temperature range for heat assimilation with respect to zooplankton, considering their rather high lethal limits. This appears to be so from the results of studies with a trophic series of Polish lakes, including one

Table 26. *Zooplankton (herbivores) production and turnover rate (P : B) over a trophic series of lakes and with heat added*

Lake	Trophic state	Production (kcal m^{-2} day^{-1})	P : B
Naroch	Mesotrophic	0.12	0.06
Miastro	Mesotrophic – eutrophic	0.30	0.08
Batorin	Eutrophic	0.71	0.18
Mikorzynskie	Eutrophic – unheated	0.62	0.11
Lichenskie	Eutrophic – heated	1.38	0.24

Source: Patalas (1970).

that received heated water from a power plant (see Chapter 7). Whereas zooplankton production and turnover rate increased with eutrophication, more than another doubling occurred when heat was added to the eutrophic lake over that without heat (Table 26). As was noted in Chapter 7, however, the heated lake was dominated by blue-green algae, which would tend to negate the benefit of high production. Zooplankton may not be the most critical part of the aquatic community when considering all the effects of heated water discharges.

Study questions

1. Could the following two situations exist in the presence of low concentrations of soluble phosphorus, the limiting nutrient?

	productivity mg C m^{-3} day^{-1}	biomass mg m^{-3}	μ day^{-1}
A	200	50	1.6
B	50	200	0.2

   Briefly explain the cycling process that would allow A to out-produce B.

2. Note the following facts from two reservoirs:

	A	B
chl a μg l^{-1}	5	10
Total P μg l^{-1}	60	30
Photic-zone depth, m	2	8
Thermocline depth, m	10	10
Secchi-disk depth, m	1	4

   Will phosphorus reduction in the inflow to reservoir A benefit "lake quality"? Explain. (Hint: Why is biomass in A less than in B in spite of higher P concentration in A than in B? Assume the C:N:P ratio shows P to be most limiting in A and B.)

3. Explain why carbon is not regarded as an important controlling nutrient for the biomass level but is apparently important for determining the relative dominance by blue-green algae in the phytoplankton? (Hint: Consider the equilibrium of inorganic carbon forms.)

4. Should in-lake eutrophication control techniques, such as PO$_4$ complexation, *precede* or *follow* external controls on P and why?

5. Of 23 lakes from which P was diverted, the monitored response of lake P concentration in over 80% showed good agreement with the Vollenweider model, but less than 50% of the 23

changed in trophic state from eutrophic to mesotrophic or oligotrophic. Explain why that could occur assuming that P is the most limiting nutrient.

6. Control of the effects of eutrophication could be effective through increased grazing of phytoplankton by zooplankton. In order for that to succeed, what changes in the phytoplankton and fish populations would be needed and why?

7. Explain why the failure of an eutrophic lake phytoplankton community to proceed to the normal dominance by blue-green algae would mean more food for plankton-eating salmonid fishes?

8. Why did Vollenweider's later model for critical P loading ($L_c = 200 \, (\bar{Z}_\rho)^{0.5}$) provide a more reasonable fit for observed data than his first attempt ($L_c \simeq 50 \, \bar{Z}^{0.6}$)? Give a real or ficticious example.

9. The following information exists for Swamp Lake:

P loading $\qquad$ $\bar{Z}$ $\qquad$ ρ $\qquad$ $\bar{P}$
300 mg P m^{-2} yr^{-1} $\qquad$ 3.0 m $\qquad$ 3.0 yr^{-1} $\qquad$ 45 μg l^{-1}

What would you say is the trophic state of this lake? Does all the information agree and why? See question 8 for a useful equation.

10. Dr. Procarbon demonstrated with in situ bioassays that the addition of NaHCO$_3$ stimulated productivity in a matter of several hours in an eutrophic lake. He therefore suggested that 95% of the carbon should be removed in the proposed upgrading of the sewage treatment facility that had been discharging secondary treated sewage to the lake. On the other hand, Mr. Anticarbon suggested that in spite of the bioassay results only phosphorus removal is necessary. Which person would you agree with and why?

11. Scum Lake is dimictic and eutrophic and is surrounded by irrate citizens who want "something done" about the lake's trophic state. The mean depth is 10 m, flushing rate is 1 per year, P loading is 1 g m^{-2} per year, and the retention coefficient is 0.5. Because the hypolimnion goes anaerobic, large amounts of P are regenerated from the sediments. Alum treatment is recommended for the lake to reduce that internal source. Would you expect the lake to retain the improvements from an alum treatment long after that first year? Why? Hint: Consider Dillon's equation $L \, (1 - R)/\bar{Z} \rho = \bar{P}$.

12. Nitrogen tends to limit growth more in eutrophic lakes than in oligotrophic lakes because
    a. the N/P ratio can be lowered by the introduction of sewage effluent
    b. the N/P ratio could be lowered if the rate of denitrification is high
    c. nitrogen fixers dominate in eutrophic lakes
    d. two of the above
    e. all of the above

13. Surface hydraulic loading ($\bar{Z} \rho$) in m yr^{-1}
    a. decreases phosphorus retention if increased
    b. increases phosphorus retention if increased
    c. has little influence on phosphorus because it indicates only water movement
    d. has little value because it cannot be directly measured

14. With respect to oxygen resources, eutrophication usually results in
    a. a decrease in the oxygen deficit rate in mass area^{-1} time^{-1}
    b. an increase in the oxygen deficit rate in mass area^{-1} time^{-1}
    c. an oxygen concentration in the hypolimnion that drops below 2-3 mg l^{-1}
    d. an increased oxygen deficit rate in mass area^{-1} time^{-1} that is a function of only autochthonous organic production

15. Dominance of the summer phytoplankton in eutrophic lakes by blue-green algae has been hypothesized to be caused by
    a. selection of the blue-greens by filter-feeding zooplankton, which stimulates growth rate of the blue-greens
    b. the low sinking rates, or even buoyancy, of the blue-greens, allowing them to receive more light than diatoms during stratified periods
    c. increased free CO_2 levels brought about by increased pH during periods of intense photosynthesis in fertile waters
    d. selective toxicity to blue-green algae of high levels of silica—with increased production, the silica level is reduced and blue-greens can prosper

16. Plankton algae effectively limit their own growth by
    a. extracting CO_2 from the water, which results in lowered pH and reduced CO_2 concentration
    b. extracting CO_2 from the water, which results in raised pH and reduced free CO_2 concentration

   c. extracting CO_2 from the water, which results in lowered C_T concentration below the limiting level

   d. extracting CO_2 from the water, which raises pH and decreases C_T and results in lower free CO_2 concentration

17. To assert that the critical depth (Z_{cr}) in most lakes is greater than the lake depth means that

   a. the calculated depth above which gross production exceeds respiration for a mixing plankton cell is greater than the lake depth

   b. the calculated compensation depth must approach the total depth of the lake

   c. the ratio of photic-zone depth to lake depth is at least 0.5

   d. plankton blooms should occur in winter as well as in spring and summer in most lakes

18. Increased temperature (daily mean) from 20 to 30 °C in the epilimnion of a lake 30 m deep should result in a

   a. temporary increase in primary production following the Q_{10} rule and a greater dominance by blue-green algal species

   b. permanent increase in primary production following the Q_{10} rule and a greater dominance by blue-green algal species

   c. temporary increase in primary production following the Q_{10} rule and a greater dominance by green algal species

   d. permanent increase in primary production following the Q_{10} rule and a greater dominance by green algal species

   (Hint: Consider the lake to be nutrient limited in summer.)

19. Dominance of the plankton by blue-green algae has been hypothesized to be caused by all of the following except

   a. a differential sinking rate between blue-greens and diatoms during periods of water-column stability

   b. a higher optimum temperature and lower K_s for a limiting nutrient for blue-greens than for diatoms

   c. a food selection by zooplankton grazers for colonial and filamentous phytoplankton

   d. an exhaustion of silica by diatoms in highly fertilized lakes

20. A lake contains 30 μg l^{-1} of chl a and the depth of visibility averages 1 m during the summer. The external P loading is 0.5 g m^{-2} yr^{-1}, the flushing rate is 3.0 yr^{-1}, and the mean depth is 4.0 m. Which of the following most likely explains the state of the lake as

a. eutrophic and caused largely by external loading that is in excess of the critical (excessive) loading estimate
b. mesotrophic and caused largely by external loading that is less than the critical loading estimate
c. eutrophic and caused largely by internal loading because external loading is less than the critical loading estimate
d. mesotrophic and caused largely by internal loading because external loading is less than the critical loading estimate

9

PERIPHYTON

Periphyton, both plant and animal, attach to submersed substrate in standing or running water. The composition of periphyton in running water can be composed of any one or a mixture of diatoms (e.g., *Navicula, Diatoma, Cymbella, Cocconeis, Synedra,* and *Ceratoneis*); filamentous blue-green algae (e.g., *Oscillatoria* and *Lyngbya*); filamentous green algae (e.g., *Cladophora, Tetraspora,* and *Stigeoclonium*); filamentous bacteria (e.g., *Sphaerotilus*) or fungi; protozoans (e.g., *Stentor, Carchesium, Vorticella* [stalked ciliates]); rotifers (not too common in unpolluted environments); and some insects (e.g., immature midges, blackflies, and mayflies).

Periphyton appear as very slick, nearly invisible coverings on rocks in fast-flowing streams or long, weaving strands of green filamentous algae or filamentous bacteria and fungi (sometimes covered with diatoms) that completely cover the substrate in a slow stream or along the shore of a lake. The latter is, or course, typical of highly enriched environments and the nuisance effects can be as great as from unwanted blooms of phytoplankton. As an example of nuisance proportions, Wezernak et al. (1974) have mapped the extent of the nuisance filamentous alga *Cladophora* (blanket weed) along the Lake Ontario shoreline by means of remote sensing. A nearshore strip 350 m wide was 66% covered from Niagara to Rochester, New York, amounting to about 15 700 kg km^{-1}. East of Rochester the strip was 277 m wide and 79% covered, amounting to 26 000 kg km^{-1}. Waves break the strands loose from their substrata and deposit windrows of decaying plant material on the beach in such situations. As such, the resulting nuisance condition can be much greater than that caused by phytoplankton.

Contribution to primary productivity

Periphyton can represent a large percentage of the primary production in fresh water. It usually contributes a more significant fraction in flowing than in standing water. Some streams may be heterotrophic in nature and receive their organic matter from allochthonous sources.

Standing water

Periphyton productivity comprises a relatively small fraction of production in deep lakes with only a small area of shallow littoral. Production depends on the percentage of area where shallowness is sufficient to allow enough light to strike the bottom. If macrophytes are abundant, the zone of periphyton growth can be larger. Shallow lakes with large littoral areas and the intertidal areas of the ocean and estuaries produce considerable periphyton. It grows attached to rooted plants, logs, and rocks. Shifting sand beaches produce little periphyton because a stable substrate is lacking.

Running water

When current velocity is appreciable, plankton are not abundant and, consequently, nearly all primary production is from periphyton. However, slow, deep rivers have generally more plankton production than periphyton. The greater depth generally causes light to severely limit periphyton growth. However, if too turbid, plankton will also be limited and the river will tend to be heterotrophic in nature.

Rickert et al. (1977) incorporated these factors in suggesting a hypothetical scheme to account for the distribution of periphyton among different habitats (Figure 103). They show that periphyton dominate the biomass in high-velocity, shallow rivers and can also be important in sluggishly moving or standing shallow waters.

Wetzel (1964) compared the three sources of primary production in a large (44 ha), shallow, saline lake in California (Table 27). In the shallow area, periphyton exceeded phytoplankton and macrophyte productivity. But overall, phytoplankton was still greatest because of the larger pelagic area. This is probably typical for most shallow lakes. Although not the largest in most cases, the periphyton contribution is still significant.

Wetzel (1975, p. 414) cites other examples showing that the contribution of periphyton to production in lentic waters can range from 1% in an oligotrophic lake without macrophytes to 62% (the dominant producer) in a shallow, rapidly flushed lake where phytoplankton production is controlled by washout.

Streams have often been considered less productive than lakes. However, McConnell and Sigler (1959) studied a 32 ha area in the upper

Figure 103. Conceptual diagram relating algal biomass and dominant algal types to water-detention time and light penetration. Shallow implies that the euphotic zone extends to the bottom; deep implies the euphotic zone does not reach the bottom. (Rickert et al. 1977)

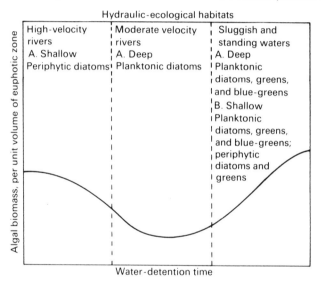

Table 27. *Comparison of the three sources of productivity in Borax Lake, California*

	Annual mean (mg C m⁻² day⁻¹)	Total annual mean (kg C lake⁻¹ yr⁻¹)
Phytoplankton	249.3	101.0
Periphyton	731.5	75.5
Macrophytes	76.5	1.4
	1,057.3	177.9

Source: Wetzel (1964).

canyon of the Logan River, Utah, which is a beautiful, fast-moving, productive trout stream, and measured rates of periphyton productivity at 1050 mg C m^{-2} per day or 334 per year. Considering the 300 mg C m^{-2} per day limit for a eutrophic state in a lake, this stream exceeds that rate by 3.5 times in the upper canyon and 17 times in the lower river. Thus, relatively shallow (less than the photic-zone depth), swift streams can produce large quantities of organic matter per unit area and still not take on the nuisance conditions of many eutrophic waters.

Methods of measurement

Standing crop can be determined by scraping an area from natural or artificial substrata and analyzing for wet weight, dry weight, ash-free dry weight, pigment content, or cell and/or species counts. The available time and specific purpose usually dictate the extent of analysis.

Accumulation or growth on artificial substrates, such as Plexiglas slides, glass microscope slides, wood shingles, and concrete blocks, can be collected and analyzed for any of the above indices. The general procedure for using artificial substrates is to stagger incubation times for certain slides to get a curve of accumulation as shown below for collections made at 2 wk, 4 wk, 6 wk, 8 wk. The time for maximum accumulation depends on the growth and loss rates. Once determined for a particular stream and season, that incubation period can be used exclusively to save analytical time:

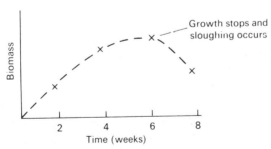

The disadvantages of artificial substrata are: (1) the species that normally occurs on the stream bottom may not be selected; (2) the accumulation rate is not productivity per se because growth starts from a bare area in contrast to what is proceeding on the normal stream bottom, which means that a much greater turnover rate is probably sustained than would occur in the existing biomass on natural substrata.

Advantages of artificial substrata are: (1) standardized, readily comparable substrata (a known area) on which organisms at each station have an equal chance for attachment and growth; (2) a precise, comparable, "rate of growth" or "rate of accumulation" can be determined; (3) data collection is easy; and (4) they are a sensitive index of water quality and effects are integrated over time.

Productivity and respiration rates can also be determined on periphyton by a light and dark bell jar using O_2 evolution or ^{14}C uptake (Vollenweider 1969b) or the O_2 curve method of Odum (1956).

Factors affecting growth of periphyton
Temperature

The same principles pertaining to the effect of temperature on phytoplankton growth also apply to periphyton growth, but some added examples may be useful. McIntire and Phinney (1965) showed that the short-term metabolism of organisms is greatly affected by temperature in an artificial stream periphyton community. The following changes occurred in respiration rates in response to abrupt (5 hr) temperature changes:

Temperature change	*Metabolism change*
6.5 → 16.5 °C	41 → 132 mg O_2 m^{-2} hr^{-1}
17.5 → 9.4 °C	105 → 63 mg O_2 m^{-2} hr^{-1}

The photosynthetic rate increased accordingly in response to an 8-hr temperature rise at a 20 000 lux light intensity (1966 footcandles):

$$11.9 \rightarrow 20 \text{ °C} \qquad 335 \rightarrow 447 \text{ mg } O_2 \text{ m}^{-2} \text{ hr}^{-1}$$

There was no effect on photosynthesis below 11 000 lux (1000 footcandles), which is about 7% of average full sunlight. Therefore, one can conclude that for a short-term response an increase in temperature at high light intensity could result in an increase in production. In this case the Q_{10} for photosynthesis was about 1.7. However, the Q_{10} for respiration was about 2.0 (Figure 104). This can be calculated according to:

$$Q_{10} = \left(\frac{K_2}{K_1}\right)^{10/t_1 - t_2} = \left(\frac{110 \text{ mg } O_2 \text{ m}^{-2} \text{ hr}^{-1}}{56 \text{ mg } O_2 \text{ m}^{-2} \text{ hr}^{-1}}\right)^{10/10} = 1.96$$

At low light intensities an increase in temperature would cause a disproportionate increase in respiration over that of photosynthesis and net production would decrease in the short term.

What can be expected in the long term from a temperature increase?

As with the plankton, the community composition would change through succession, tending toward blue-green algae and lowered diversity. Coutant (1966) showed in one of the first studies of effects from power-plant discharge that temperature at times reached 40 °C (104 °F) in the Delaware River. When the temperature exceeded about 30 °C (86 °F), blue-green algae were favored, which would result in greater light-saturated rates of photosynthesis and thus greater rates of net production. These shifts in community composition and diversity are shown in Figures 105 and 106. Coutant also noted that periphytic biomass was 8 times greater and chl *a* 2.5 times greater in the heated channel (38–40 °C) versus normal Columbia River water, where the maximum daily mean reaches 21 °C.

Light

As with the phytoplankton, adaptation in the periphytic algae occurs so that the direct effect of light is modified by high and low light adapted communities ("sun" and "shade"). This is important because shaded conditions along streams are common and the extent to which adaptation can occur determines how much of a possible increase in enrichment could effectively be utilized. McIntire and Phinney (1965) showed very interesting physiological and compositional shifts in their artificial stream community in response to changing light intensities (Figure 107). The principal effect was a higher rate of photosynthesis at low light intensity but a lower light-saturated rate for the shade-adapted community.

Percentages refer to the fraction of community that was diatoms (D),

Figure 104. Relation of respiration and temperature, showing example rates at a 10° interval. (Phinney and McIntire 1965)

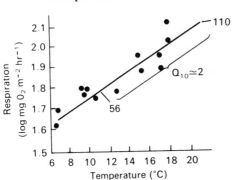

blue-green algae (B-G), and green algae (G). Five hundred footcandles equals about 5360 lux or 0.032 langleys per min. In spite of the overall slower growth in shade-adapted conditions, biomass accumulated on substrate will attain levels similar to that at high light but will require more time. This is shown in Table 28 from the same artificial stream. Biomass in the light stream reached an apparent saturated level in two-thirds the time necessary for the shade stream.

The interaction of low light and nutrient enrichment is worth mentioning at this time in relation to McIntire's work. The effect of shade adaptation is that nutrient enrichment may not increase the photosynthetic rate. McIntire and Phinney (1965) showed that there was a response to CO_2 increase in the light-adapted community but not in the shaded community (Figure 108).

The explanation for this phenomenon seems to be that if photosynthesis and CO_2 removal is great enough and the CO_2 level (or any other nutrient) is low enough to limit growth, the light-adapted community

Figure 105. Results of periphyton collection from artificial substrata over a 7-day period ending June 4, 1959, in the Delaware River at Martin's Creek Plant. (After Trembly 1960 and Coutant 1966)

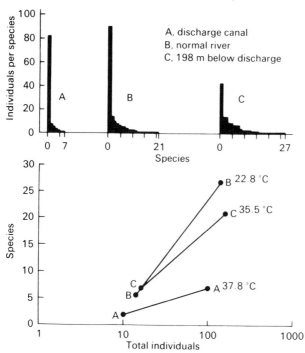

will cause CO_2 to limit, and increases in CO_2 will stimulate a further increase in photosynthesis. Because the pH change was only 1.45 units, it was discounted as directly influential in that case. However, instances may exist in which nutrient content is low enough to provoke

Figure 106. Comparison of microorganisms in warm water and normal Delaware River water at Martin's Creek Plant. (After Trembly 1960; Coutant 1966)

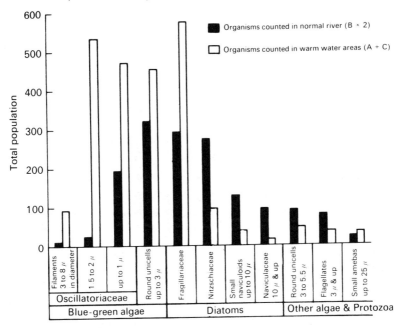

Table 28. *Biomass accumulated in mg per slide surface for artificial stream communities held in light (550 footcandles) and shade (200 footcandles) conditions (values are maxima)*

Duration	Light	Shade
180 hr	140	5.4
2 wk	120	89
6.5 mo	593	—
9 mo	—	565

Source: McIntire and Phinney (1965).

limitation in shade-adapted communities so that increased enrichment would have a stimulatory effect.
The interaction of light and temperature can have a dominant influ-

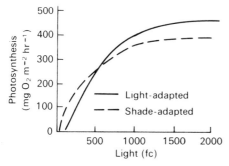

Figure 107. Effect of light on shade-adapted (200 fc) and light-adapted (550 fc) communities composed of diatoms (D), blue-green algae (B/G), and green algae (G). The light-adapted community consisted of 46% D, 42% B/G, and 12% G. The shade-adapted community consisted of 67% D, 26% B/G, and 7% G. (McIntire and Phinney 1965)

Figure 108. Light- and shade-adapted periphyton community response to increased light over a gradient of CO_2 content. (McIntire and Phinney 1965)

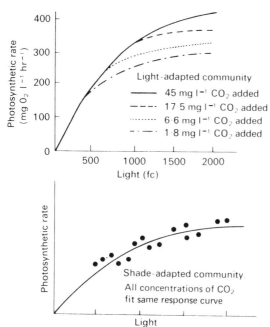

ence on the seasonal occurrence of any given species. Alteration of either factor causes shifts in species composition, as was shown with phytoplankton. An example of the interaction of light and temperature in producing the seasonal changes in *Cladophora glomerata*, a nuisance filamentous green algae, in Lake Erie was demonstrated by Storr and Sweeney (1971). Optimum growth occurred at 18 °C (64 °F) and growth stopped at 25 °C (77 °F) for this alga. From natural photoperiod and temperatures along with laboratory growth data related to photoperiod and temperature, a close comparison (Figure 109) between observed and estimated biomass was found. In spite of suboptimum temperature and photoperiod in October, which resulted in low estimated biomass, observed biomass increased greatly, probably in response to increased nutrient content following lake overturn. Such an increased periphyton growth in the autumn following lake turnover and PO_4 increase was also observed in Lake Sammamish (Porath 1976). That too was in spite of decreasing light conditions. Thus, one must conclude that such a close fit between periphyton growth and the physical factors light and temperature holds for a rather stable nutrient regime, but where limiting conditions occur, fluctuations in nutrient levels can dominate growth response.

Turbidity from suspended sediment can be a serious limitation to productivity by reducing light penetration. The action of sediment can also scour the substrata, reducing the biomass of periphyton. Such

Figure 109. Observed biomass of *Cladophora* in Lake Erie and estimated levels based on experimental light and temperature response. Observed values for temperature given in degrees Fahrenheit. (Storr and Sweeney 1971)

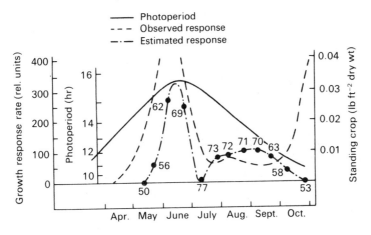

environments actually show up as heterotrophic streams on analysis of the oxygen balance with P:R < 1 (see Chapter 1).

Current velocity

The community types, or groupings of periphytic species, can be selected by velocity alone as shown by McIntire (1966) for the two experimental velocities. At a high velocity of 38 cm per sec, a feltlike growth was produced composed largely of diatoms. At the much reduced velocity of 9 cm per sec, long oscillating filaments of the green algae *Stigeoclonium, Oedogonium,* and *Tribonema* developed. Obviously, the erosion rate is too high for the filamentous forms at high velocity.

Sphaerotilus growth in artificial streams has been shown to increase with velocity (Figure 110). Likewise, Schumacher and Whitford (1965) showed that the uptake rate of ^{32}P by *Spirogyra* increased with velocity (Table 29).

From the standpoint of biomass and production of periphyton, the accumulation of biomass is fastest in swift current, but the total biomass eventually tends to equalize at both high and low velocities. As a result, export and therefore production, as well as the turnover rate of biomass, are greatest in the fastest current. The stimulation of velocity on productivity occurs because metabolic rates are greater as current velocity increases, a result of the rapid exchange of water around cell surfaces in attached plant mass that removes wastes and

Figure 110. Effect of current velocity on *Sphaerotilus* growth. In all cases sucrose was added at 5 mg l^{-1}; and velocity is indicated at right in centimeters per second. (Modified from Phaup and Gannon 1967)

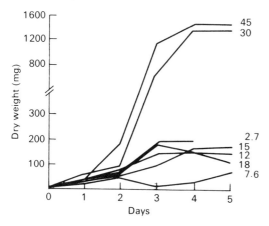

replenishes the nutrient content. Schumacher and Whitford concluded that the significant effect of velocity between 1 or 2 cm per sec and about 45 cm per sec is due to the increased rate of diffusion at the plant surface. A more efficient utilization of nutrients by the cell can therefore occur because the diffusion gradient between water and the cell surface is maintained.

Production, turnover rate, and export of periphytic mass should all increase in proportion to velocity. The observed rate of accumulation on artificial substrates should also increase with velocity as long as scouring is not severe. A maximum velocity, above which loss through erosion becomes greater than accumulation through growth, is probably around 50 cm per sec (Horner 1978).

Nutrients and velocity

In contrast to the phytoplankton in lakes, little is really known about critical concentrations of nutrients that either relate to growth rate or cause nuisance growth in the periphyton. Because of the stationary nature of periphyton, it would be logical that higher concentrations of a limiting nutrient would be necessary in the surrounding media to produce the same growth rate in periphyton as in phytoplankton in a lentic environment. However, as velocity increases, the necessary concentration should decrease.

Because production is proportional to nutrient supply rate (nutrient concentration × velocity), the ultimate biomass accumulation at moderate velocities (20–40 cm per sec, no scouring) should also be proportional to nutrient supply rate. The result is the same whether the mechanism is the increased mass load or maintenance of the concentration gradient through increased turbulence. An increase in velocity, or in limiting nutrient concentration, or both, should increase the

Table 29. *Effect of velocity over a range of low rates on the uptake of* ^{32}P *by* Spirogyra

Velocity (cm sec^{-1})	Uptake (counts min^{-1} g dry weight^{-1})
0	78 424
1	92 748
2	98 602
4	158 861

Source: Schumacher and Whitford (1965).

accumulation rate. Either an increase in concentration at low velocities (<10 cm per sec) or an increased current velocity should have the effect of increasing periphyton growth in environments of low current velocity. However, biomass accrual may be greater with nutrient increase than velocity increase if the concentration is below the growth rate saturation level. In swift-moving streams (>40 cm per sec), on the other hand, increased concentration (if much above *growth rate saturation levels*) should not result in a marked biomass or production increase. Unfortunately, such a growth rate saturation level in streams is unknown. Although the evidence for this is not great, there is some indication in the work with *Sphaerotilus* by Phaup and Gannon (1967). They found that the concentration of sucrose required for maximum growth is inversely related to velocity. More recently Horner (1978) has found that diatom accrual is enhanced by both velocity and limiting nutrient concentration increase and that above about 50 μg l^{-1} PO$_4$-P stimulation is greatest by velocity increase, but below that level nutrient increase enhances accrual. Velocity increase at low nutrient concentration reduces accrual.

Inorganic nutrients (N, P, and C)

Enrichment of nearshore areas has been shown to cause nuisance growths of *Cladophora* (blanketweed) in the Great Lakes (Lakes Huron and Erie). Nuisance growth is associated with waste effluents and runoff from agricultural activities. However, good light penetration in unenriched areas also allows biomass accumulation, apparently without enrichment. Plants in unenriched areas with good light can grow to four times the depth they grow in waste-affected areas, which indicates an interaction between light and nutrients (Neil and Owen 1964).

Experimental results to determine the causative factor in the nuisance growth showed that P was the most important nutrient. N, P, and organic fertilizer were added in various combinations to nearshore areas with no existing *Cladophora* growth and were seeded with rocks containing a start of the periphytic alga. The experimental period extended from June 24 to September 25. Results, illustrated in Figure 111, essentially showed that by July 19 only a small growth existed from sheep-manure input and by August 8 a bed of *Cladophora* had developed from P, N, and K.

Significant growth occurred neither with N alone nor at the control site. An important point in this experiment is that periphyton growth in standing water is most apt to be limited by nutrients at relatively high concentrations because current is low (see earlier comments on streams). In this case it is PO_4-P, and the rate of 2.7 kg per day to that affected area would amount to a renewed concentration each day of about 250 μg l^{-1}. That rate of input represents a waste load from about 1000 people, thus a very heavily loaded area.

This nuisance species has also been shown to increase its cell storage of PO_4 near waste sources and following storms with increased urban runoff. Algae near such nutrient sources were longer, greener, and denser (Lin 1971). Critical loading limits that relate to attached algal growth in littoral areas should be developed. Pitcairn and Hawkes (1973) have suggested an upper limit of 1.0 mg l^{-1} PO_4-P for running water, beyond which no increase in *Cladophora* biomass would occur. However, this seems very high and may apply mostly to sluggish streams. Enrichment of swift-flowing (40 cm per sec), low-nutrient water on Vancouver Island, British Columbia, increased the biomass of diatoms. Biomass increase resulted from experimental increases of PO_4-P from 3 μg l^{-1} to 9 μg l^{-1} during the period when light was not limiting. No response was shown from nitrate addition (Stockner and Shortreed 1978).

Figure 111. Effect of inorganic nutrients on *Cladophora* growth in Lake Huron. Experimental period was June 24 to September 25. (Neil and Owen 1964)

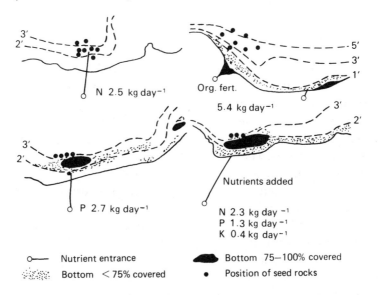

N 2.5 kg day^{-1}

Org. fert.

5.4 kg day^{-1}

P 2.7 kg day^{-1}

Nutrients added

N 2.3 kg day^{-1}
P 1.3 kg day^{-1}
K 0.4 kg day^{-1}

o—— Nutrient entrance

Bottom 75–100% covered

Bottom < 75% covered

• Position of seed rocks

Enrichment may have no effect on periphyton growth in swift-moving streams if nutrient content is above growth-rate limiting levels as pointed out earlier. The data from Darby Creek (Patrick 1966), a eutrophic stream with existing nutrient concentrations of 0.83 mg l^{-1} P and 1.7 mg l^{-1} N, showed no effect on biomass from nutrient additions varied one at a time to produce levels up to 28 mg l^{-1} NO$_3$-N, 5 mg l^{-1} PO$_4$-P, 15 mg l^{-1} NH$_3$-N, and 0.8 mg l^{-1} glucose. Observations indicated that there was no change in abundance of periphytic diatoms or succession to blue-green algae, but that some shift in diatom species occurred with increases in the following:

NO$_3$-N, *Melosira varians* was favored

N and P, *Synedra parasitica* increased

Glucose, *Gomphonema parvulum* and *Nitzschia* increased

P, *Synedra rumpens* increased

Whether such species change was definitely a function of those nutrients is questionable. The main point, however, is support for the previously mentioned contention that nutrient increases above the growth-rate limiting levels, particularly at high velocities, will usually have little or no effect on biomass and production. The effect of increased velocity and/or nutrient concentration (even if above the growth-rate limiting level) will be most pronounced when normal velocity is low. If velocity is high, however, an effect will occur only if the limiting nutrient concentration is below the critical level, possibly around 50 μg l^{-1} for PO$_4$-P.

Organic nutrients

Dissolved organic nutrients favor bacterial slimes dominated by filamentous bacteria (*Sphaerotilus*) and fungi. "Bacterial slimes" are aggregates of *Sphaerotilus, Zooglea,* and fungi. *Sphaerotilus* may have a different appearance morphologically under varying environmental conditions. For example, *S. natans* may appear similar to *Cladothrix dichotoma* and in turn to *Leptothrix ochracea.* These organisms are obligate organotrophs. They can create nuisance slime conditions, usually in running water, where a source of dissolved organic substances is discharged. They have long been problems near untreated waste discharges from pulp mills.

The culture-determined nutrient requirements for *Sphaerotilus* are mono and disaccharide sugars at a concentration of 20 to 50 mg l^{-1} as well as organic acids, amino acids, and inorganic N and P. Their presence or absence, however, seems definitely related to the presence of sugars and the concentration is rather critical.

Phaup and Gannon (1967) determined the effects of sucrose and velocity on production and community type in experimental 210 m outdoor channels. The periphyton community, developed on strings used as artificial substrata in river water without additional enrichment, included: *S. natans,* the algae *Melosira, Navicula, Cosmarium, Euglena,* the protozoans *Tetrahymena, Colpedium,*amoeba, and some insects, tendipedids (midges), and simulids (blackflies). Sucrose enrichment increased production as evidenced by increased mass accumulation on the string substrata. Growth was maximum within 30 hr with 5 mg l^{-1} sucrose and was proportional to water velocity over a range from 18 to 45 cm per sec, as noted earlier (Figure 110), at a temperature of 20 to 28 °C. The river water contained 0.5 mg l^{-1} N and 0.001 mg l^{-1} P. The maximum yield of biomass was not obtained below 17 °C because *S. natans* was replaced by another filamentous bacterium that was much less prolific. This finding has bearing on the winter phenomenon in which *S. natans* often covers more stream area than it does in summer. The cause for this is related to temperature and metabolic rate. At high temperature, utilization and growth are rapid, resulting in the quick depletion of nutrients, and, consequently, the minimum required organic content occurs over a smaller area than in winter when low temperatures exist.

In a similar experimental channel study, Ormerod et al. (1966) showed that succession of species within the bacterial slime community was affected by a combination of organic and inorganic nutrients. Bacterial slimes replaced algae when SSL (spent sulfite liquor) was added but were only partly replaced when sewage was added, because the additional inorganic nutrients favored algae. Two streams were studied with the same concentration of SSL (0.1 ml l^{-1}). The following organisms dominated under two different stream conditions:

Soft-water stream: *Fusarium aquaeductum*

Hard-water stream: *Sphaerotilus natans*

Under variable temperature with velocity of 5 cm per sec, the following change in the periphyton community occurred:

SSL addition at 0.1 ml l^{-1}: Algae → *Sphaerotilus*

SSL + sewage at 0.5–1.0 ml l^{-1}: Algae increased with some *Fusarium*

At a constant temperature of 12 °C and a water velocity of 15 cm per sec, the following community changes were observed (Figure 112):

Sucrose addition at 10 mg l^{-1}: Algae → *Sphaerotilus* and *Zooglea*

Sucrose (10 mg l^{-1}) + PO$_4$-P at 2 µg l^{-1} and up to 1000 µg l^{-1}: Algae → *Sphaerotilus* → *Fusarium*

The addition of other nutrients, such as N, K, Na, Mg, Ca, Fe, and vitamin B_{12}, had no effect upon succession. P seemed to be the critical inorganic nutrient. Of course, an increase in temperature resulted in an increase in growth rate (Figure 113) but the pattern of ultimate dominance did not change – only the rate of succession. Regardless of temperature, the addition of P favored dominance by *Fusarium*. This is most interesting because it shows that although heterotrophic microorganisms are favored by highly reduced dissolved organic matter, the species actually dominating can be strongly influenced by inorganic nutrients.

Figure 112. Effect of phosphate on slime community growth. Channels contained 10 mg l^{-1} sucrose and 1.9 mg l^{-1} NH$_4$Cl. (Omerod et al. 1966)

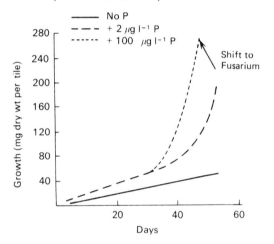

Figure 113. Effect of temperature on slime community growth. Channels contained 10 mg l^{-1} sucrose, 0.19 mg NH$_4$Cl l^{-1}, and 0.34 mg K$_2$HPO$_4$ l^{-1}. (Omerod et al. 1966)

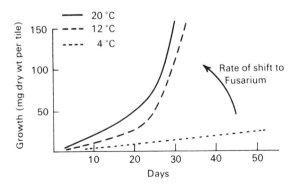

The concentration of total dissolved organic matter is probably of less importance than its composition. If diverse, nuisance conditions may not develop. Cummins et al. (1972) added leaf leachate, a natural source of allochthonous organic enrichment, to an artificial stream at 10 times the natural concentration, up to 30 mg l^{-1} DOC, which is a level similar to that received by heavily polluted streams. In 10 days 85% of the DOC was utilized and no change in the diversity of macroinvertebrates resulted. Moreover, a diverse fungal flora developed without dominance by *Sphaerotilus,* which probably would have occurred if 30 mg l^{-1} of DOC as sucrose had been added.

Other indications that a diverse DOC source will not lead to a dominance by nuisance heterotrophic populations was shown by Ehrlich and Slack (1969). Figure 114 shows the accumulation of periphyton biomass in two artificial streams; one received inorganic N and the other a yeast extract at 50 mg l^{-1}. The increase in algal biomass was greatest with the addition of inorganic N because preliminary mineralization of the N was not required. With the yeast extract, an intermediate step was required for bacterial mineralization, which then lead to algae. Although blue-green algae dominated in the organic N stream, filamentous heterotrophic bacteria like *Sphaerotilus* did not.

Figure 114. Biomass of periphyton in two artificial streams, one receiving inorganic and the other organic (yeast extract) N. Diatoms and green algae were dominant at the start of the experiment in both streams, bacteria peaked early in stream with organic N, but at the end, green algae dominated the inorganic stream and blue-greens dominated the organic stream. (Ehrlich and Slack 1969)

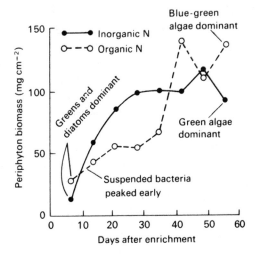

Based on the previous evidence from artificial-stream studies, one can conclude that the principal cause for nuisance heterotrophic growth (*Sphaerotilus*) is the presence of highly degradable, dissolved organic matter, particularly sugar, and specifically:

1. Wastes containing sugar in the range of 5–30 mg l^{-1} leads to nuisance growth dominated by *Sphaerotilus*.
2. Sugar plus inorganic nutrient (mainly P) enrichment leads first to *Sphaerotilus* then to a domination by fungi.
3. Diverse DOC compounds even up to 30 mg l^{-1} (such as natural leaf leachate) leads to a variety of nonfilamentous (and therefore nonnuisance) heterotrophs and algae.

An explanation for the presence or absence of nuisance slimes is not always possible from the usual measurements of BOD and/or total dissolved organic carbon. Curtis and Harrington (1971) stress that heavy slime outbreaks have been noted where very low concentrations of BOD exist as well as no slime development in spite of existing high BOD levels. They note that the former could be due to the presence of rapidly degradable substances, such as sugar, whereas the latter could be due to the presence of inhibitors. In any event, secondary treatment that removes 90% of the BOD in waste water, with such highly degradable substances as sugar effectively eliminated, will usually solve the problem of nuisance growth in streams.

Periphyton community change as an index of waste type

Although organic waste such as sewage can result in an increase in periphyton biomass with mono species dominance, certain species are typically characteristic of various levels of organic matter, and a characteristic transition occurs among successional stages during natural degradation. In this regard Fjerdingstad (1964) has described a successional response of the periphyton community to a gradient of sewage enrichment (Table 30). This successional response is similar to the Saprobien System first described by Kolkwitz and Marsson (1908) and often referred to as the process of "stream purification."

The length or presence of these zones in a stream depends upon flow rate (dilution) and temperature. These factors, in turn, determine the dispersed concentration, its contact time, and, therefore, the biological uptake rate. A high flow rate and low temperature results in a low concentration and low uptake, so that a lesser effect is stretched over a longer stream distance. With low flow and high temperature, the reverse occurs.

Sládečková and Sládeček (1963) have described the succession of periphyton in a series of Czechoslovakian reservoirs polluted with sugar-beet waste and sewage. The combined effect of these wastes showed typical zones, which are illustrated with drawings (Figure 115). Three representative periphyton communities in the reservoirs, receiving various amounts and types of waste effuents (sugar beet and sewage), were collected on glass slides. The communities indicate different degrees of water quality change: 51 – relatively "clean" water conditions (oligosaprobity); 52 – "recovery" conditions (beta-mesosaprobity); 53 – "polluted" conditions (alpha-mesosaprobity).

Table 30. *Water quality characteristics and the associated response of the microorganism community to organic waste*

Zone	Chemical	Biological
Oligosaprobity (clean water)	BOD < 3 mg l^{-1} O_2 high Mineralization of organic matter is complete	Diatoms diverse Filamentous green algae present Filamentous bacteria scarce Ciliated protozoans scarce
Polysaprobity (septic)	H_2S high O_2 low NH_3 high	Algae present but not abundant Protozoa absent Bacteria abundant – fecal, saprobic, and filamentous[b]
α Mesosaprobity (polluted)	Amino acids high H_2S low-none $O_2 < 50\%$ saturated BOD > 10 mg l^{-1}	Algae scarce – some tolerant forms[a] Filamentous bacteria abundant[b] Ciliated protozoans abundant[c] Few species – biomass great
β Mesosaprobity (recovery)	$NO_3 > NO_2 > NH_3$ $O_2 > 50\%$ saturated BOD < 10 mg l^{-1}	Diatoms not diverse – biomass great[d] Ciliated protozoans only present[e] Blue-green algae abundant[f] Filamentous green algae abundant[g]
Oligosaprobity (clean water)	Stream recovered or "purified"	

[a] *Gomphonema, Nitzschia, Oscillatoria, Phormidium, Stigeoclonium* often dominate. [b] *Sphaerotilus, Zooglea, Beggiatoa.* [c] *Colpidium, Glaucoma, Paramecium, Carchesium, Vorticella.* [d] *Melosira, Gomphonema, Nitzschia, Cocconeis.* [e] *Stentor.* [f] *Phormidium, Oscillatoria.* [g] *Cladophora, Stigeoclonium, Ulothrix.*
Source: Fjerdingstad (1964); Sládečková and Sládeček (1963).

Note the lack of diatoms and preponderance of ciliated protozoans and filamentous bacteria in the "polluted" water community (53). According to this system of evaluating water-quality change, the various species are grouped according to their position with respect to distance or travel time from a waste outfall and, therefore, to the amount of "self-purification" that has taken place. The occurrence of these groups of species or communities, then, is considered indicative of various degrees of organic "pollution."

Figure 115. Illustrated pattern of periphyton response to waste from three Czechoslovakian reservoirs. See text for explanation. (From Sládečková and Sládeček 1963)

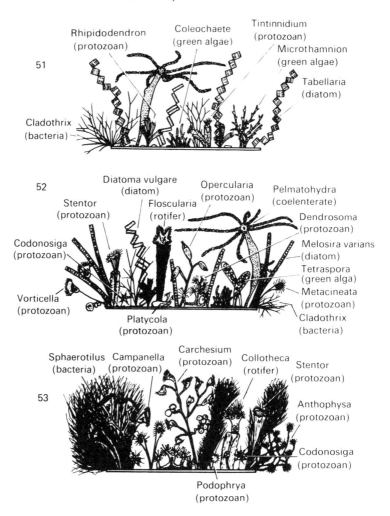

Often the index chl a : total organic matter (or C) determined on artificial substrata can be used to determine the area and degree of effect of organic waste on periphyton (Grzenda and Ball 1968). This ratio was found to lie between 10 and 17 in clean-water areas and 2 and 3 in areas receiving sewage effluent. Thus, heterotrophs occupy a greater portion of the biomass as organic waste increases. In that case a pigment index was used instead of chl a. However, the objective with such a measurement is to be able to determine the relative change in the autotrophic and heterotrophic fractions. This can be done by microscopic examination if the unaffected station has a high proportion of autotrophic forms, in which case the chl a : C ratio could be designated "autotrophic" and affected station measurements compared to that, recognizing the normal variability in the chl a : C ratio in algae.

Effect of toxicants

The effect of toxicants on periphyton communities has not been adequately studied, but toxicants nonetheless show unique effects. Usually the total biomass is reduced from an inhibitor, and the tolerant species become dominant and possibly increase in abundance. Predators may be eliminated from a slug dose, followed by rapid growth increase in algal periphyton as a result of no grazing once the toxicant has dispersed. For example, about a week after spraying with DDT to control forest insects, forest streams have become intensely green. The addition of Zn^{2+} to artificial streams showed an increased reduction of the number of algal species over a concentration range of 0-9 mg l^{-1} (Williams and Mount 1965). Noteworthy in this experiment was the elimination of *Cladophora* at the lowest concentration tested, $- 1$ mg l^{-1}. In this instance water was taken from a holding pond that also supplied a high concentration of particulate food; consequently, *Sphaerotilus* and fungal slime organisms also increased as Zn^{2+} increased and algal species decreased. Without the dead organisms, little increase in decomposer biomass would have occurred.

Of course, Cu^{2+} is another heavy metal to which algae are very sensitive. $CuSO_4$ is a common algicide. Hynes (1960, p. 81) cited work by Butcher showing the effects of Cu^{2+} in industrial waste in an English stream (Table 31). Recovery of species diversity and biomass (33 000 mm^{-2}) occurred 8 km downstream from the point of discharge.

The question of nuisance

Under proper conditions of light and temperature, which usually set limits on potential growth, enrichment with either inorganic

or organic nutrients promotes the dominance of certain prolific periphyton species and thus an increase in production. This is often considered undesirable or a "nuisance" condition for the following reasons:

1. A secondary BOD is created that can deplete O_2 downstream
   as the filaments break off, float away, and decompose.
2. The clogging of water intakes with floating clumps of filaments
   occurs.
3. Undesirable taste and odors can be created if the affected
   stream is used for a water supply.
4. The dense mats can cover the bottom, restricting intragravel
   water flow, which can inhibit fish reproduction.
5. The presence of dense mats and long stringy filaments can
   interfere with fishing and recreation.
6. The mats and filamentous clumps reduce the habitat of benthic
   animals and can cause direct physical damage.

Although these problems seem to prove that the excessive growths are
a "nuisance," the increased production could be interpreted as a possible benefit under some circumstances because it could lead to increased production of fish food and fish. An example of the benefits
were indicated by Warren et al. (1964) as a result of their studies in
Berry Creek, Oregon, where sucrose added at concentrations from 1-4
mg l^{-1} produced mats of *Sphaerotilus* and dense populations of midges.
The diversity of both periphytic microscopic and macroscopic organisms decreased. Table 32 shows the results of annual biomass levels,
fish-food consumption, and production in energy units in enriched
and unenriched sections of Berry Creek.

The production of cutthroat trout appears to have been benefited in
this instance from the enrichment, possibly in spite of the mats of

Table 31. *The effects of Cu waste on stream periphyton collected on
artificial slides*

Upstream station	Periphyton density 1000 mm^{-2} 3 wks^{-1}
Cu^{2+} low	*Stigeoclonium, Nitzschia, Gomphonema, Chaemosiphon,* and *Cocconeis* were most abundant
Downstream from waste	Density 150-200 mm^{-2} 3 wks^{-1}
Cu^{2+} 1.0 mg l^{-1}	*Chlorococcum, Achnanthes*

Source: Hynes (1960).

Sphaerotilus that developed. If you were a waste discharger and believed (as do also some nonproducers of waste) that a goal for sound water-quality management is the use but not the abuse of ecosystems, you might argue that this level of enrichment was beneficial to Berry Creek. However, the following points must be considered by the thoughtful manager:

1. The difficulty of "fine tuning" (regulation of input to allow "benefits" without "detriments") a variety of ecosystems receiving a variety of wastes.
2. An unacceptable time delay in recovery if systems become overloaded.
3. The costs and problems of monitoring to provide data for fine tuning.
4. The instability risks of intense management. That is, a fine-tuned system may be highly susceptible to environmental changes. For example, trout are known to grow very fast in highly productive areas, but with associated temperature and pH, NH_3, and O_2 conditions that may be otherwise marginal.

Table 32. *Insect biomass, fish-food consumption, and fish (cutthroat trout) production in artificially enriched (1–4 mg l^{-1} sucrose) and unenriched sections of Berry Creek, Oregon (values are totals for a portion of each year)*

	Annual values (K cal m^{-2})				
	Unenriched (U) dark	Unenriched (U) light	Enriched (E) dark	Enriched (E) light	E:U
Insect biomass					
1960–61	2.19	5.03	20.4	12.2	4.5
Fish-food consumption					
1961	8.38	6.06	20.15	15.64	
1962	9.46	8.09	19.07	21.55	
1963	7.45	8.36	13.05	9.00	
mean	8.43	7.50	17.42	15.40	2.1
Fish production					
1961	−0.21	0.01	2.13	2.20	
1962	0.49	−0.07	3.70	4.80	
1963	0.58	0.99	2.51	1.65	
mean	0.29	0.31	2.78	2.88	6.3

Source: Modified from Warren et al. (1964).

Slight changes in weather, such as cloudy or hot weather, could cause a total elimination of the highly prized, fat trout. In the case of Berry Creek, the trout were planted fish, and whether or not reproduction was successful in the enriched portions of the stream was not determined. As noted in the introduction, increased production may be a trade-off for decreased stability.

Study questions

1. What is the probable order in the rate of periphyton accrual in the following streams where phosphorus is the limiting nutrient and give a reason for the placement of each stream.

Stream	Average velocity (cm sec⁻¹)	Average PO_4-P (μg l^{-1})
1	10	50
2	70	50
3	40	50

2. Test streams A and B are similar except that velocity in A is 30 cm sec⁻¹ and in B 5 cm sec⁻¹. If organic carbon, nitrate, and phosphate were added such that the concentrations upon mixing were 30 mg l⁻¹ DOC, 50 μg l⁻¹ PO_4-P, and 500 μg l⁻¹ NO_3-N, in which stream would the greatest periphyton production occur and what would the order of dominance be between filamentous heterotrophs and algae with stream distance and why?

3. A dense accumulation of a gray, feathery, growth in running water usually indicates:
   a. inorganic nutrient enrichment to greater than growth rate limiting levels
   b. velocities in the range of 10–50 cm sec⁻¹
   c. organic carbon enrichment to levels of about 10 mg l⁻¹ or above
   d. a and b
   e. b and c

4. Accumulation of a dense biomass of the filamentous green alga *Cladophora* on bottom substrate in streams
   a. occurs in an oligotrophic stream if velocity is increased within the tolerable range
   b. is favored if velocity is increased within the tolerable range;

but increased enrichment with inorganic nutrients, especially P, is required if the stream is oligotrophic

c. is favored if velocity is increased within the tolerable range; but increased enrichment with dissolved organic carbon is required if the stream is oligotrophic

d. occurs with increased enrichment with inorganic nutrients, especially P, and velocity changes have little effect

5. An increase in the heterotrophic:autotrophic index downstream from an input of untreated sewage affluent could result from all of the following except

a. temperature increase with BOD load and flow volume constant

b. BOD load increase with flow and temperature constant

c. flow volume increase with BOD load and temperature constant

d. flow volume decrease with BOD load and temperature constant

6. If phosphorus concentration is decreased 50% in the following two streams the response in terms of decreased periphyton accrual will be

Stream	Average velocity (cm sec^{-1})	PO_4-P (μg l^{-1})
A	40	50
B	10	50

a. greater in B than in A because P is more efficiently used at low velocity

b. greater in A than in B because more P is required to overcome erosion at high than low velocity

c. greater in A than in B because P is more efficiently used at high than low velocity

d. greater in B than in A because 40 cm sec^{-1} in too great for appreciable accrual to occur

10

MACROPHYTES

The higher plants in aquatic environments are usually rooted with rigid cell walls, and asexual and sexual reproduction (flowers). Asexual, or vegetative, reproduction is usually more effective. Most of those considered here are submergent types, occupying either littoral zones of lakes or the slow-moving or stagnant reaches of rivers. Emergent types dominate in the very shallow and marshy areas, but they do not present the nuisance problems provoked by submergents.

Habitats
Running water

Most rooted macrophytes are not able to withstand too great a current and an adequate sediment deposit must exist for their rooting. Light is generally satisfactory, because rivers are usually shallow, but turbidity may still limit in some instances. Butcher has listed five types of waters, based on current and sediment type, that influence nutrient availability and in which the associated turbidity affects light penetration (Butcher 1933; Hynes 1960). The types with representative plants are:

1. Torrential on rock or shingle – this habitat contains mostly mosses (*Fontinalis*); larger macrophytes are not stable in the high velocities.
2. Nonsilted on shingle – a "sluggish" stream; milfoil (*Myriophyllum*) and water crowfoot (*Ranunculus*).
3. Partly silted on gravel and sand – this habitat, because of bottom type and turbidity, has, for example, *Ranunculus*, pondweeds (*Potamogeton*), and arrowweed (*Sagittaria*).

4. Silted on silt - the sedimented bottom and increased turbidity favors pondweeds (*Potamogeton*), water lilies (*Nuphar*), and the aquarium plant (*Elodea*).

5. Littoral - very little current with mud substrate. Emergent vegetation is common, for example, reeds and bulrushes.

Standing water

The distribution of macrophyte development in standing water is dependent on slope in the littoral region. The more gradual the slope, the greater the potential to collect sediment, as well as the more area available to light penetration, and thus optimal growth. Spence (1967) has observed in Scottish lochs that macrophyte diversity and abundance are favored by brown muds and an alkalinity greater than 50 mg l^{-1} as $CaCO_3$. Vegetation is usually absent from rocky or sandy beaches.

The characteristic zones in the littoral region are:

1. Emergent vegetation zone: grasses, rushes, and sedges, which depend on lake sediments for nutrients but otherwise carry on photosynthesis in the atmosphere, which supplies CO_2.

2. Floating leaf plant zone: water lillies, some pondweeds, and duckweed (*Lemna*), most of which obtain their nutrients primarily from sediments but carry on some photosynthesis in direct contact with the atmosphere.

3. Submerged vegetation zone: pondweeds, hornwort (*Ceratophyllum*), naiads (*Najas*), milfoil, arrowweed, and stonewort (*Chara*), which obtain nutrients from both soil and water and are dependent on CO_2 in water for photosynthesis.

Significance of macrophytes

Macrophytes accelerate lake aging. As plants develop on enriched sediment in the increasing littoral areas, they continue to collect more sediment around them from extermal sources as well as from their own death. They are a very effective trap and therefore accelerate the filling in of a lake.

Macrophytes act as a physical substrate for periphyton and insects and thus provide a very positive benefit to the ecosystem, facilitating overall food-chain production as well as contributing to primary production. Wetzel and Hough (1973) argue that because most lakes in the world are small and shallow macrophytes contribute a significant portion of the primary production in fresh water (see Wetzel 1975). Also, they are involved in the life history of fish, supplying surfaces for egg incubation and protection of young. For example, small northern pike are very cannibalistic and without weed beds for protection few survive. However, if too dense, weeds can overprotect juvenile fish from predation, which leads to stunting in both predator and prey species.

Probably the greatest significance of macrophytes to the public is the problem of nuisance growths. Dense biomass interferes with water use by clogging water intakes, impairing esthetics and recreation, and depleting the O_2 resources. A special problem with *Myriophyllum spicatum* (Eurasian water milfoil) has developed in the United States and Canada. The problem has progressed from the East through the Midwest and since the mid-1970s has become pronounced in British Columbia and Washington State. The plant has been a problem since the early 1960s in the Midwest. In the Tennessee River impoundments 2, 4-D has been applied at 67 500 kg (150 000 lb) per year over an area of 1500 ha (3700 A) to control the plant (Smith et al. 1967). Two other exotic species that are nuisances are hydrilla and water hyacinth. Those occur in the Southeast. Macrophyte control alternatives will be discussed later.

Response to environmental factors

In order to understand why macrophytes cause problems in some environments and not others and to recommend sensible controls for their excessive production, some points about their response to natural environmental factors must be understood.

Light

The quality of light plays a similar role with macrophytes as with phytoplankton. Red light is more favorable to production than other wavelengths (Figure 116). Hodgson and Otto (1963) showed that the longer wavelengths toward the reds produced the greatest growth, although growth was substantial in the blue and green portion.

Macrophytes are high-light lovers, and their abundance and distribution, therefore, can vary greatly depending on light quantity. The zone of maximum growth is limited usually to only a few meters because the attenuation of light with depth is great in most waters and maximum growth has been shown to occur at light intensities between 100% and 75% of full sunlight.

As shown in Figure 117, *Elodea canadensis* has a high-light optimum. The measurements of ^{14}C assimilation rate were made on a sunny day

Figure 116. Growth of *Potamogeton pectinatus* during a 4-wk exposure to three wavelength ranges. (Hodgson and Otto 1963)

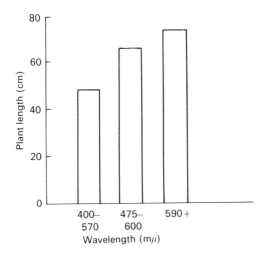

Figure 117. Photosynthetic rates of *Elodea canadensis* suspended in pickle jars at the indicated percentages of incident light in a shallow pond. (Modified from Hartman and Brown 1967)

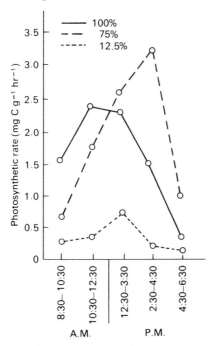

with a maximum intensity of about 2700 footcandles (28 000 lux). The optimum intensity for this species is probably near full sunlight. A photosynthetic curve for intensities between 75% and 100% would have likely peaked around noon. The result indicates the extent that macrophyte growth can be affected by the reduced penetration of available light at depths greater than 1 to 2 m in turbid reservoirs and streams and the consequently significant limiting effect.

Because of their sensitivity to light availability, the distribution and abundance of macrophytes can be affected to a large extent by climate and water clarity. For example, the year-to-year variation in plant abundance in Pickwick Reservoir, Alabama, as related to incident light and rainfall are shown in Figure 118.

These data indicate that the spring growth period was relatively dry and sunny in the 2 yr in which the growth of rooted aquatic plants (*Najas* sp.) reached nuisance proportions. Nuisance proportions in this case means that about 2400 of the 20 000 ha (50 000 A) reservoir was covered with submergent plants that reached the surface, allowing mats of periphyton to develop and quiet pools to harbor mosquito larvae.

The sunny, dry springs at Pickwick Reservoir indicate that more

Figure 118. Mean daily incident light and mean monthly rainfall recorded at Florence, Alabama, during the "critical" period of plant growth in Pickwick Reservoir from April 15 to May 31 during 1961 through 1967. (Peltier and Welch 1970)

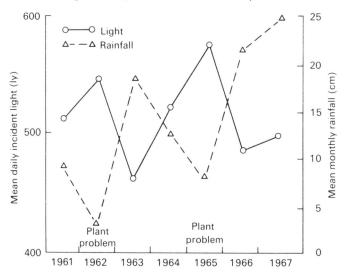

incident light and less turbidity actually made more light available to the plants, the factor that caused the variation in abundance. The depth of 10% light intensity in this reservoir in spring ranged from 1 to 2 m and the depth of 1% intensity from 2.5 to 4 m. Turbid conditions could therefore greatly affect the area suitable for plant growth.

Robel (1961) also found turbidity to be a dominant factor in controlling the growth of *P. pectinatus* in a Utah marsh (Figure 119). Such turbidity, which can also be caused by plankton algae, contributes greatly to the limitations of growth of submersed macrophytes in hypereutrophic systems. Phillips et al. (1978) suggest that the cause for macrophyte decline is more complex and is related first to the shading produced by a coating of epiphytes that grow profusely in rich waters.

Temperature

Temperature acts as a control on light-saturated rates of photosynthesis and approximates the Q_{10} rule, as was shown with plankton and periphytic algae. Field observations of *P. pectinatus* have shown the maximum growth rate to occur in the spring, coincident with the greatest increase in light while temperature remained rather low (10 °C). Hodgson and Otto (1963) state that the pondweed can develop at temperatures from 10–28 °C. The interpretation must be that the greatest rate of growth occurs in young plants, but that this rate increases as temperature increases up to its thermal optimum, because light-

Figure 119. Relation between *P. pectinatus* biomass and turbidity in a Utah marsh. (Robel 1961)

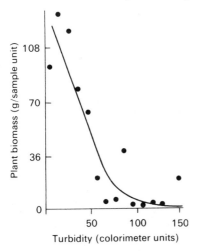

saturated rates of photosynthesis increase with temperature. Thus, if the river had been heated during the spring when light increased, then growth would have been greater; a temperature increase from heated-water discharges would probably select species with temperature optima matching the new conditions, as was the case with algae. Hynes (1960) observed that *Vallisneria spiralis* (wild celery) is usually a warm-water aquarium plant in Britain and has been found in the natural environment only where heated effluents occur. However, nuisance macrophyte growth related to heated-water discharges has not been a recognized problem.

Nutrients

Although changing light availability resulting from decreased amounts of plankton algae or suspended sediment may be responsible for causing permanent changes in macrophyte abundance, nutrient enrichment changes are most often blamed. To evaluate that point it is necessary to determine which nutrients are important for macrophytes and whether their most significant source is sediment via the roots or water via the leaves. Although N, P, and C are the significant macronutrients for macrophytes, as with algae (K may also be significant), experimental work shows that the growth of rooted macrophytes may not be affected by high levels of N and P in the water (for example, 500 and 50 $\mu g\ l^{-1}$, respectively) as are phytoplankton and periphyton. Experiments with *Potamogeton pectinatus* were conducted in aquaria with running water over a gradient of nutrient concentrations in the inflow. There was no significant difference in growth (stem elongation) over a sizable range in levels (Table 33). This suggests that growth is

Table 33. *Growth (stem elongation) of* P. pectinatus *in aquaria with continuous flow*

	NO$_3$-N (mg l^{-1})	PO$_4$-P (mg l^{-1})	22-day growth (mm) Sediment	Sand
Full nutrient	1.71	0.26	359	310
Half nutrient	0.89	0.11	303	264
Tap water	0.44	0.03	351	288
Tap water with sealed roots	0.58	0.02	303	84

Source: Peltier and Welch (1969).

either saturated at very low concentrations in the water or that most of the nutrient comes from sediment via the roots. When roots were sealed in sand from overlying water, growth was significantly less than it had been when the roots were in rich river sediment, the source of the original culture. The same striking effect of sediment and minor effect of high levels of nutrients was shown by Mulligan and Barnowski (1969) with the nuisance plant *Myriophyllum spicatum* (Table 34). The greater effect of sediment-derived nutrients is particularly evident in those results. There is some limitation at very low water concentrations (5 μg l^{-1} P, 50 μg l^{-1} N) with roots in sand. With sediment, the effect is quickly swamped out.

Somewhat in contrast, Nichols and Keeney (1976a, 1976b) found that foliar uptake of N exceeded that of roots if the concentration of NH_3-N in water was 100 μg l^{-1} or more. Such concentrations were present in spring and fall in Lake Wingra (Wisconsin) water and also in storm water inputs, but in summer concentrations were very low. At low concentrations in the water plants were just as successful as with high NH_3 content, and in fact plant N content in the lake was related to N content of sediments, suggesting that overall sediment N is most important.

As a point of interest, the nutrient that is most limiting to the plant in the natural environment is determined most readily by tissue content. Critical tissue concentrations have been determined in many species experimentally by relating tissue content versus growth in plants exposed to high concentrations of a complete medium of elements (Figure 120). Tissue analyses of naturally occurring specimens have

Table 34. *Growth (biomass dry weight) of* Myriophyllum spicatum *under different treatments of sediment type and water nutrient concentration*

Water nutrient (mg l^{-1})	Percent sand/percent sediment			
	100/0	67/33	33/67	0/100
0.5 P; 5.0 N	1.43	—	—	—
0.05 P; 0.5 N	1.26	—	—	—
0.005 P; 0.05 N	0.68	3.78	4.51	5.63

Source: Modified after Mulligan and Barnowski (1969).

shown very few plant populations to be nutrient limited, however (Gerloff and Krombholz 1966). An exception was recently reported by Gerloff (1975). He showed that potassium (K) was limiting to *M. spi-catum* in shallow, eutrophic Lake Wingra.

A workable hypothesis to explain the nutrient question is that organic sediment stimulates the growth of rooted plants regardless of nutrient content in water and that the organic content, as well as soil texture, also affects the species composition of rooted macrophytes. It may be that organic content controls the availability of N and P.

In some early but important experimental work, Misra (1938) observed a species selection by sediments of different humus content in Lake Windermere, England. This followed the hypothesis by Pearsall (1920, 1929) that distribution was controlled by the substratum. Three types of sediment were noted in the lake and each had a characteristic plant assemblage (Table 35). Interestingly enough, experiments in jars showed *P. perfoliatus* to grow best on the sediment type in which it

Figure 120. The critical tissue concentrations of N and P (concentrations allowing maximum growth, dashed line) in algae-free cultures of several species of macrophytes were consistently around 1.3% N and 0.13% P. The above results are with *Vallisneria americana*. (Modified from Gerloff and Krombholz 1966)

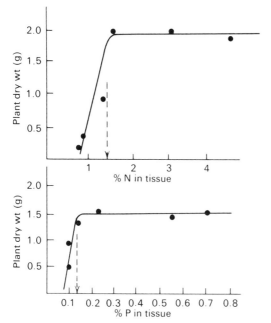

was found in the lake. Subsequent laboratory experiments showed that each dominant species similarly grew best in the sediment type where it was naturally found.

Experimental work has shown that *P. pectinatus* growth is greater in organic sediment than in sand by 15%. Growth was 400% greater in organic sediment than in sand when roots were sealed from overlying water (Table 33). Similar results were found with *Najas*, in which growth increased linearly with increased lake sediment in spite of constant or decreasing nutrient levels. There was some indication that the quantity of acid-soluble P in sediment was important (Figure 121, Martin et al. 1969). The work with *M. spicatum* mentioned earlier is also in agreement (Table 33). Gessner (1959) cites other work with *Elodea* and *Myriophyllum* species showing comparable findings.

The mechanism that explains the effect of increasing organic content in sediment on nutrition of rooted macrophytes is not well known. One possibility is that ion exchange capacity increases in organic sediment. Misra (1938) found that Fe and H exchange increased with organic content (7% to 42%) and Ca exchange was greatest in an intermediate range of organic content. The antagonizing ions tend to be adsorbed onto the organics. Another possibility is that nutrient supply (particularly P) is no doubt greater as organic content increases. Reducing conditions are more likely because the bacterial respiration

Table 35. *Association of plant and sediment types in Lake Windermere, England, and 1 month's growth of* P. perfoliatus

Mud type	Dominant plant species	Percent humus	Dry weight in mg
Inorganic course brown silt	*Isoetes* (quill wort)	8.04	467
Moderately organic black flocculent mud	*Potamogeton perfoliatus*	12.26	778
Highly organic brown mud	*Sparganium minimum* and *Potamogeton alpinus*	24.00	298

Source: Misra (1938) with permission of Blackwell Scientific Publications, Ltd.

and binding sites that remove P from solution are apt to be taken up by organic material, thus allowing P to be more mobile and result in high interstitial PO_4-P concentrations.

As a general rule, the increased enrichment of water with N and P no doubt favors macrophytes indirectly if the slope in the littoral region is gentle enough for the settlement and accumulation of the organic matter and nutrients resulting from increased plankton production. Other favorable factors include adequate light penetration and a gentle current that will allow some sediment to build up so that plants can root and maintain themselves. Otherwise, nutrients in the water are probably of secondary importance for the development of rooted macrophytes. Their abundance and distribution is most likely controlled by average light intensity and the organic content or other nutrient and texture conditions of sediments. Thus changes in available light because of depth and turbidity changes and the respective enrichment of water and, subsequently, sediment will greatly influence the composition of the producer groups. Wetzel and Hough (1973) have suggested a generalized sequence among the various types of primary producers and increased eutrophication (Figure 122). Note the early dominance by phytoplankton and periphyton followed by macrophytes

Figure 121. Effect of increasing fractions of lake-bed sediment (mixed with sand) on the yield of *Najas* and the measured concentrations of nutrients in experimental containers. (Martin et al. 1969)

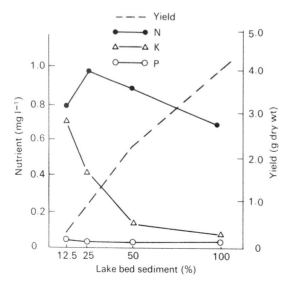

(submersed), which later decline because of light limitation by plankton algae or periphyton. The periphyton should be the first indicator of eutrophication because it grows nearest the source of external nutrient input, thereby contacting relatively high concentrations of nutrients. Phytoplankton would be expected to precede macrophytes in response to a gradual increase in external loading, possibly showing an even more pronounced separation in time than indicated in Figure 121. That would be a result of the time necessary to accumulate a rich sediment, which has been shown to favor macrophytes. Periphyton would show a greater rate of increase than phytoplankton as eutrophication progressed because self-shading would not be as limiting to periphyton and they would probably respond more to the higher nutrient concentrations. The decline of macrophytes is probably caused by shading by periphyton as well as by phytoplankton.

The problem of *M. spicatum* deserves special consideration because of the rapidity in which it dominates a system. Much of its success is related to its efficient vegetative reproduction. Sections of the plant break off easily, float away, and become rooted in a new area. These fragments can also be transported to uninfested systems, which is probably how the species spread from the East Coast to the West Coast of North America in less than 15 yr. In some areas it has been observed to decline to 10–20% or less of its former level following several years of dominance (Elser 1967; Wile 1975; Carpenter and Adams 1978). Cause(s) for such a decline has not been identified.

Although the plant is an exotic species, there must be definable environmental reasons why it so quickly outcompetes other species

Figure 122. A generalized scheme for the changes in submersed macrophytes, attached algae, and phytoplankton in lentic environments. (Modified from Wetzel and Hough 1973)

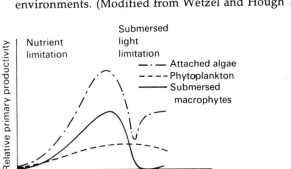

and totally dominates an environment. Coffey and McNabb (1974) observed that it grew better under the ice at low light intensity than other species in a Michigan lake. It therefore had a larger biomass at the beginning of the year than other species and dominated through a shading effect. *M. spicatum* has taken over during the past several years in Union Bay, Lake Washington, in spite of increased clarity from declining phytoplankton densities. Although that does not tend to support the low-light tolerance idea, it has nevertheless been almost the only plant to grow in some of the turbid Tennessee River impoundments. The plant can reach the surface up to a depth of about 3 m and then extend out horizontally on the surface, which would be an effective light limiter to plants below.

A final point on the problem species *M. spicatum* comes from Hutchinson (1970b), who suggests the possibility that its competitive ability may be derived from the basic chemical character of the water. Hutchinson (1970b) showed from the work of others in Sweden that *M. spicatum* seemed to prefer the higher pH and higher Ca content relative to two other species (Figure 123). He thought *M. spicatum's* success may have been related to its ability to use HCO_3. Spence (1967) also noted the presence of *M. spicatum* in lakes with HCO_3 concentrations ≥ 60 mg l^{-1} ($CaCO_3$). Increased fertility tends to raise pH and shift the carbon equilibrium to more HCO_3 and less CO_2, because of increased utilization of CO_2 by plankton, as has been pointed out earlier. In this

Figure 123. Distribution of three species of *Myriophyllum* in Sweden as a function of pH and water hardness. (Modified from Hutchinson 1970b)

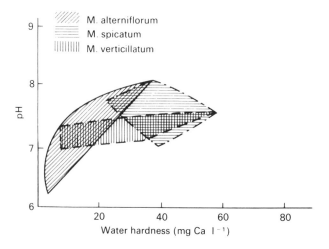

way eutrophication may be contributing to the accelerated growth and rapid dominance of *M. spicatum* in lakes and reservoirs. However, its introduction and take-over has also occurred in rather oligotrophic environments, so the above mechanism cannot be a total explanation.

Controls for macrophytes

Harvesting vs. herbicides

Several techniques are available to control or provide temporary relief from excessive macrophyte biomass. These have been reviewed by Nichols (1974) and will only be briefly mentioned here with one or two examples. It should be stressed that for the most part these techniques treat only the symptoms of the problem and not the cause, which is usually excessive nutrient load.

Probably the most logical aspect of harvesting is the direct removal of nutrients from the lake. The nutrient removal aspect can be significant, but it depends on the external supply. Wile (1975) showed that 1.8×10^6 kg of plant biomass harvested from a 265 ha area of Lake Chemung accounted for 170 kg P and 1780 kg N or 20% of the P retained in the lake and 11% of the external supply (0.17 g m^{-2} per year). Carpenter and Adams (1978) have shown that harvesting milfoil in Lake Wingra, Wisconsin, could remove as much as 37% of the external P income. Because the plants obtain most of their nutrients from the sediment, this means that the sediment pool could be depleted ultimately and that the internal P source could be reduced. Harvesting *Elodea densa* in Long Lake, Washington, could remove 60% of the external P loading and 41% of the total (external plus internal) loading (Welch et al. 1979). Wile (1979) has subsequently reported that continued harvesting of 19% of the area in Lake Chemung, Ontario, from 1972-8 removed 47% of the external P loading per yr. At that rate all the available P could theoretically be removed from the sediment in about ten yr.

Nichols and Cottam (1972) and Nichols (1974) found that two or three cuttings significantly reduced (from 50% to virtual elimination) the *Myriophyllum* crop in Lake Mendota, and the effect persisted through the following year. Even one harvest held the plant height to one-half uncut areas after 2 mo of regrowth. Overall, Nichols (1974) found that harvesting costs were similar to that for herbicide treatment, ranging from about \$38 to \$225 ha^{-1} (\$15 to \$90 A^{-1}). Bryan et al. (1975) reported costs of \$150 to \$400 ha^{-1} (\$60-\$160 a^{-1}). Newroth (1975) has criticized harvesting as giving only temporary relief from nuisance problems in the Okanogan Lakes in British Columbia, Canada.

Although the initial capital costs may be greater with harvesting than with herbicide treatment, the similarity of effectiveness and overall costs over the long term would seem to favor harvesting when other advantages are considered (Wile 1975):

1. An immediate reduction is achieved without using foreign substances

2. Biomass is removed, eliminating its contribution of detritus that would decompose, releasing dissolved nutrients and pose a possible O_2 problem

3. The effect of plant "pumping" of sediment nutrients is diminished

4. The effect on animals is minimized by not totally removing all plants

The herbicide most commonly used to control macrophytes is 2, 4-D. It is most effective in granular form at a rate of about 112 kg ha^{-1} (100 lb A^{-1}), and the cost is around $125 ha^{-1} ($50 A^{-1}) (personal communication, Dr. D. R. Sanders, U.S. Corps of Engineers). The cost for diquat, another commonly used herbicide, is about double that for 2, 4-D. Newroth (1975) reports that costs for diquat range from $250 to $750 ha^{-1} (see Dunst et al. 1974 for other herbicides and rates). Although 2, 4-D is toxic to animals, a rate of application of 50 kg ha^{-1} in liquid form did not show toxic effects in the food web where the application was in an area of low dilution (Wojtalik et al. 1971). The 2, 4-D remained in water for 2 wk but in the plankton for 6 mo, although there was little accumulation by plankton-eating fish. Residues in sediment were gone within 3 mo. In areas of significant dilution the material is relatively short-lived. Therefore, opposition to 2, 4-D based on its side effects of toxicity is not strongly supported by observations. However, the natural selection of stronger strains of particular nuisance species (or of different, stronger species) by continued treatment could develop, as in the case of fish and insects with DDT, although such an effect with herbicides has not yet been demonstrated. Also, if natural parasites of nuisance plants exist, such as fungi or bacteria, herbicidal treatment could prevent their natural action. A "disease" is suspected to have caused the near elimination of milfoil from Chesapeake Bay in the 1960s (Elser 1967).

Drawdown

Nichols (1974) stressed that drawdown can be quite inexpensive and effective if facilities exist, such as in reservoirs. He cited results from three separate studies in which reductions ranged from 50% to 90% in the area covered. The cause for the reduction was thought to be due to freezing and the desiccation of plants and roots. Drawdown in winter, therefore, was considered most effective.

Another effect of drawdown is increased aeration and consolidation of sediments. Organic content should decrease with increased oxidation if aeration and drying are effective. This could adversely affect the nutrition of plants following refilling, considering the role of organic sediment previously discussed, as well as deepen the littoral area, which could possibly reduce light availability.

Deepening from consolidation would result because the weight of overburden would force water out, thus compacting the soil particles closer together (Smith et al. 1972). As the amount of organic matter increases, the overburden pressure would decrease and consolidation would be less effective. Fox et al. (1977) have shown substantial consolidation of lake sediments from drying with the subsequent poor growth success of most macrophytes but little reduction in organic content.

Sediment blankets

The application of sand or gravel onto existing sediments to a thickness of 15–20 cm has been used to control macrophytes (Nichols 1974). Theoretically, growth should decline because of decreased nutrition to the roots, which has been shown previously in the laboratory when natural, organic sediments were mixed with sand (Tables 33, 34 and Figure 120). In a survey of macrophyte distribution in Scottish lochs, Spence (1967) observed that rock, gravel, or sand substrata were only sparsely colonized by macrophytes or not at all. Although logical, the experimental application of plastic sheeting and/or sand and gravel has shown mixed results (Born et al. 1973b). Whereas control was effective the first year, there was more growth on blanketed areas the second year. Nevertheless, the technique is applicable to small areas, particularly because the cost per area is relatively high. Application in each problem situation should be approached cautiously, however, until the effect has been demonstrated.

Biological control

The grass carp, or white amur, has been used successfully to control aquatic vegetation. Considerable doubt exists, however, about its alleged inability to reproduce in temperate climates, which would allow effective control of its abundance and distribution. Therefore, the use of the grass carp is discouraged and even forbidden in some states. Accepted use must await a better understanding of its reproductive potential in temperate climates or the proven feasibility of using unisex populations.

Other potential biological controls include several pathogens and insects. None of these have been recommended for widespread use, however.

Dredging

Removal of lake sediment by suction dredging is very costly. Nichols (1974) cites a cost of up to $1.25 m^{-3}. In 1979, the cost would probably be two to three times that. The most likely long-term effect would be a deepening of the lake, thus restricting the available light. Because problem lakes usually have large shallow areas, the amount of sediment that must be removed for effective deepening over an extensive area could be prohibitive in cost.

Sediment oxidation

Ripl (1976) has suggested the application of inorganic oxidation agents, such as $NaNO_3$, to organic sediments to reduce nutrient availability. This method would be directed more toward the cause for excessive macrophyte growth, which is apparently associated with an increased organic content of sediments. Reduction of the organic content of sediment could restrict the availability of P (Ripl 1976).

Study questions

1. Under what conditions would harvesting be a logical alternative for macrophyte control as well as nutrient control and give one other advantage and disadvantage compared to 2, 4-D applications.

2. People in an environmentally oriented community felt they were misled by consultants who told them their lake would be "cleaned up" if they would only vote for the tax increase that would pay for a sewer around the lake to intercept septic tank effluents. The algal content did in fact decrease and the lake cleared, but in three to four years macrophytes grew in profusion throughout the shallow areas, reaching the surface from

depths of 3 m. Why would you not be surprised by this chain of events?

3. Eutrophication control by P diversion may result in a trade-off of more rooted macrophytes for less plankton algae because the

   a. extinction coefficient is increased allowing macrophytes to outcompete plankton algae for more of the remaining P in the water column

   b. extinction coefficient is decreased allowing macrophytes to outcompete plankton algae for more of the remaining P in the water column

   c. extinction coefficient is increased allowing macrophytes to utilize the high interstitial P concentrations in the sediments

   d. extinction coefficient is decreased allowing macrophytes to utilize the high interstitial P concentrations in the sediments

11

BENTHIC MACROINVERTEBRATES

The benthic macroinvertebrates of standing and running waters include many orders of organisms that are of interest in their own right. The mere existence of large populations of strange-looking nymphs, naiads, and larvae of insects within the rubble and gravelly substrata of streams is fascinating. That so few people ever know of even their existence is a pity. Like any other part of the food web, they have a function.

The macroinvertebrates include grazers, detrital feeders, and predators. They process and utilize the energy entering streams from either autochthonous periphyton production or from allochthonous sources, such as leaves, needles and bark from the forest, or organic wastes from man or other animals in the watershed. Macroinvertebrate species can be lumped into feeding types, such as large-particle detritivores (shredders), small-particle detritivores (collectors), grazers (periphyton scrapers), and predators. Furthermore, the representation of the community among these groups is distinctive, as streams change typically from upstream heterotrophic to autotrophic in mid regions and ultimately back to heterotrophic in lower reaches (Cummins 1974).

In lakes or the sea they are largely dependent upon the autochthonous or allochthonous production that is sedimented to the bottom. Also, a diverse community of macroinvertebrates is instrumental in stream purification – the processing of organic matter that leads ultimately to CO_2, water, and heat. As pointed out earlier, the efficiency with which that process proceeds depends upon the diversity of the community – the more individual niches that are filled, the quicker and more completely the conversion process takes place. Overloading streams with organic or inorganic enrichment or the introduction of toxic substances will decrease the diversity and thus the efficiency of "purification."

The macroinvertebrates of fresh water are dominated by the insects and their diversity is greatest in running water. The orders Ephemeroptera (mayflies), Plecoptera (stone flies), Trichoptera (caddis flies), Diptera (true flies), and Odonata (dragonflies and damselflies) usually make up the majority of the biomass. Because of their number – 100 to 200 species is a common occurrence in running water – their identification to the species level is difficult. Other important groups in fresh water are the Mollusca (snails and clams), the Annelida (worms and leeches), and the Crustacea (scuds, sow bugs, and crayfish). In the marine environment the latter groups represent the community to the practical exclusion of the insects. In brackish water environments there is considerable overlap and some of the more tolerant freshwater species of insects can be relatively important. See Appendix A for a classification system for freshwater macroinvertebrates.

The purpose here is to treat the macroinvertebrates as indicators of the kind and relative magnitude of waste entering the system. As such, they offer the greatest advantage of any group. They are sedentary and relatively easy to sample compared to the highly motile fishes, and because their longevity is greater, fluctuations in biomass and species composition are less pronounced than in plankton. Therefore, the macroinvertebrate fauna can readily reflect changes in the input load or type of waste even when sampled as infrequently as monthly or bimonthly. Semiannual sampling is often adequate to detect significant changes.

The life history of the macroinvertebrates includes either three or four stages in the case of the insects: egg, naiad (or nymph), and adult, or egg, larvae, pupa, and adult. Although completion of more than one life cycle in a year is not uncommon, particularly with the midges, many require one or more years for completion (see Usinger 1956). The Oligochaeta, Mollusca, and Crustacae have two- to three-stage life histories and have one or more generations per year. Most of the insect's life is spent in the immature stage, with the adult, terrestrial portion usually devoted to the purpose of reproduction. Therefore, bottom substrate samples largely contain the naiads and larvae of insects, and major changes occur in the species composition through the spring and summer because emergence time varies among the populations with photoperiod and temperature. Such natural community changes can be identified, however, apart from the effect of wastes by previous study or with a representative control area. The mere presence of a variety of long-lived species that remains relatively stable in diversity and bio- mass from one month to another means that water quality has remained relatively unchanged. Any alteration in water quality, even though brief, can easily be reflected in that community. If decimated in the interim, the only replenishment source for the remainder of that life cycle is from tributaries or upstream. As indicators of water quality, along with their inherent interspecies variability in tolerance to the variety of wastes that enter aquatic ecosystems, the macroinvertebrates are extremely valuable.

Sampling for benthic macroinvertebrates
Quantitative estimates
The existing standing crop and species representation of stream invertebrates can be sampled with several types of units, including the following:
1. Surber square foot sampler most often used in shallow areas.
2. Standard area enclosure (heavy metal frame) with screen often used in riffle areas of deeper streams.
3. Hess sampler is a cylindrical unit used in deep or shallow riffles and is more efficient than the above two units.
4. Peterson and Ekman grabs are used either in deep pools in rivers or lakes, and the Ekman is restricted to small-particle substrates.

None of these samplers is representative for all habitats (riffles, pools, shorelines, etc.). The size of the area sampled must be determined by the density of the organisms so that a minimum sample size is possible with minimum sorting effort. Selectivity of certain types for different organisms must be considered. For example, the Surber selects for those organisms that wash from rocks easily and are large enough to be caught with a net. Small and newly hatched larvae and naiads are easily lost. Artificial substrates have been used with considerable success. Multiplate samplers, which are similar to slides, for periphyton are one type (Hester and Dendy 1962). Wire baskets filled with natural rubble substrate and buried at each station have been successful (Anderson and Mason 1968; Mason et al. 1970). Some species are probably missed by these methods, particularly less mobile ones. However, comparisons have shown good agreement with dredge collections (Anderson and Mason 1968). Also, more individuals are usually collected with the artificial units. Of distinct advantage is the comparison among stations by using the same substrata. Nevertheless, some analysis of the naturally occurring community structure should be included to determine what portion of the community is sampled with artificial substrates.

Qualitative estimates

Hand nets can be used in weedy areas, along mudbanks, and in debris or in places where it is difficult to use any of the above quantitative samplers. A visual search on protruding substrates such as logs, large rocks, sticks, and so on is often useful.

Although these procedures give occurrence and therefore contribute to a species list, they cannot give quantitative data on relative abundance and thus are not as useful for communicating findings and statistical analysis.

Natural factors affecting community change

Substrate and current are important factors in selecting kinds and quantity of macroinvertebrates. Macon (1974), in discussing the effect of current on macroinvertebrate community type, groups velocities as very swift (>100 cm per sec), swift (50-100 cm per sec), moderate (25-50 cm per sec), slight (10-25 cm per sec), and very slight (<10 cm per sec). Current erodes substrata composed of either rocks, stones, or gravel, which interact to determine the kinds of organisms in the community (Hynes 1960). Current determines sediment transport. The greater the current, the larger the substrate type (gravel to boulders) and the freer the interstices of fine sediments; as current decreases, deposition increases. The two types of environments, running and standing water, and associated community types can be compared.

Running water

Eroding substrata is the rule in running water, particularly in the upper reaches of streams. Species are numerous and of many groups, including worms, leeches, snails, clams, insects, and some crustaceans (scuds), but insects usually dominate the community. In such an environment the substrate is usually clean, interstices are free of sediment, and animals are adapted in characteristic ways: limpets adhere to smooth stone surface, insects have sharp claws (stone flies, caddis flies, and mayflies) to cling to stone surfaces. Some are dorso – ventrally flattened so that the whole body fits firmly against stones

(Heptageniidae). Many species, particularly stone flies, do not tolerate sediment because it fills interstices among the rocks, which hinders attachment for herbivorous feeding and restricts hiding areas. Dense algal growth on rocks tends to clog interstices, thus producing a surface unfavorable for flattened forms but favorable for midges and small mayflies. Snails and leeches attach to clean rocks. Many of these forms can tolerate velocities from 100–200 cm per sec (Macon 1974, p. 136).

As an example of how the type of eroding substrata affects density, particularly as the substrate becomes more fixed, results from English streams are significant (Table 36). The number of taxa varied from 11 to 25 but showed no special trend with substrate type. However, obvious species selection did occur within these types of eroding substrata. For example, *Rhithrogena* (mayfly) preferred stones free of vegetation, whereas *Hydropsyche* was mostly found on periphyton-covered stones. Midges comprised from 40% to 54% of the fauna on the four vegetation-covered substrate types, but only 5% to 17% on the other three. Diversity (ratio of species number to individuals) decreased considerably as density increased.

Depositing substrata, where interstices are filled because of reduced current, is typical in lower reaches of streams where sediment accumulates. This can also occur from erosion or sediment containing

Table 36. *Fauna abundance in English streams versus substrate type; fauna abundance in an unpolluted African river in summer*

Substrate type	No. m^{-2}	
English streams		
Loose stones	3 316	
Stones embedded in the bottom	4 600	
Small stones with fine gravel	3 375	
Blanket – weed on stones	44 383	
Loose moss	79 782	
Thick moss	441 941	
Pondweed on stone	243 979	
African river	*Coarse sediment*	*Stones*
Eroding substrata	6 710	4 730
Stable depositing	12 590	7 570
Unstable depositing	4 450	6 660

Source: English streams: Hynes (1960), Table 2, data from Percival and Whitehead, 1929; African river: Chutter (1969), Tables II and VI.

waste. In such an environment burrowers become dominant, for example, worms which live under the sediment surface, build tubes, and feed on detritus, and insects that burrow (midges and ephemerid mayflies). Without a clean, hard substrate to adhere to and living space among sediment-free rubble, clingers disappear. Midges, most of which are tube builders and net spinners, become abundant. Also, detrital-feeding clams occur. The abundance may not be greatly reduced in depositing substrata compared to eroding substrata. Stability of substrata may have more influence on organism density than sedimentation (Table 36). Hynes (1960, pp. 30, 31) has clearly illustrated typical representatives of the depositing and eroding substrata environments.

Standing water

Lakes are represented in general by two types of substrata, which relate to trophic state. An oligotrophic lake without an accumulation of sediment in the littoral region has clean rocks and fauna similar in some respects to that of eroded substrata in streams, for example, stone flies, but few of the burrowing types. Of course, where fine sediment prevails at depth, the community is largely midges, clams, and worms. Eutrophic lakes and reservoirs with macrovegetation and sedimented littoral regions harbor a varied fauna. In the depositing substrata community in such an environment, whether at depth or in shallows, midges usually dominate and some clams and worms occur at much higher densities than in oligotrophic lakes. A varied fauna occurs in and on the vegetation, including insects of the groups stone flies, mayflies, dragonflies, beetles, and midges.

Controls

To monitor waste discharges or to detect effects in receiving waters, a "control" must be selected for the natural variations caused by current, substrate, elevation, and temperature. Control stations must be located in areas unaffected by the waste but comparable in substrate and current to the test (waste-affected) stations. Otherwise, a significant difference in mean densities, species, or diversity between test and control cannot be interpreted as clearly a result of only the waste. If control stations are physically comparable with test stations, then interpretation of effect can be rather straightforward. If not comparable, conclusions must be qualified.

If control stations are unavailable, that is, current, substrate, and elevation at stations unaffected by waste are not similar to test stations, a control period is required before waste is introduced to receiving waters. Such a period must be sufficiently long to include normal variability. One year is usually a minimum, but two years of data is much preferred.

Because elevation difference usually causes emergence differences within insect groups, the interpretation of the presence or absence of such a group must not be mistaken for an elevation effect.

Oxygen as a factor affecting community change

Certain species and even whole groups of macroinvertebrates show various levels of tolerance to DO (dissolved oxygen). The groups in which most species are usually intolerant of low DO are the stone flies, mayflies, and caddis flies.

However, even within these groups many species can survive for a limited period of time at concentrations as low as 1 mg l⁻¹ (Macon 1974). Resistence to low DO can be partly due to morphological adaptations for maintaining a current over the animal's gills, such as with some mayflies.

At the other extreme, there are a few organisms whose special physiological and morphological capabilities allow them to tolerate or resist low DO for a considerable period. These include the air breathers: the moth fly (*Psychoda*), mosquito (*Culex*), rat-tailed maggot (*Eristalis*), pulmonate snails (*Physa* and *Limnaea*), beetles, and adult water boatman and back swimmers. Also, *Sphaerium* and *Pisidium* clams may be abundant in low-oxygen environments. Some tubificid worms and chironomid midges have hemoglobin, which can increase in oxygen-poor conditions. A high glycogen content and reduced activity allow these organisms to withstand prolonged oxygen minima (even lack) by meeting reduced metabolism with anaerobic glycolysis. Hemoglobin has a greater affinity for O₂ and probably loads up under lower O₂ tension easier than haemocyanin. The success of tubificids, particularly *Tubifex tubifex* and *Limnodrilus hoffmeisteri*, in environments that become anaerobic for extended periods, partly results from their rapid reproductive rate. They can quickly repopulate the area once the anaerobic period has passed. The same can be said for the tolerant chironomids (Brinkhurst 1965).

Those species with an intermediate tolerance to low DO include the crustaceans, *Asellus* and *Gammarus*, blackflies (*Simulium*), and some leeches (Hirudinea) and craneflies (*Tipula*). Some mayflies are adapted to low tensions with the consumption of O_2 remaining rather constant down to low levels. Figure 124 shows curves for *Cloëan* and *Leptophlebia* (Fox et al. 1935 in Macon 1974). Thus, these two species can show normal activity down to very low O_2 levels, whereas one *Baetis* species was limited at much higher levels and both *Baetis* species failed to survive at low concentrations. *Cloëan* has the morphological advantage of being able to increase circulation over its gills when DO is low.

Figure 124 illustrates the difficulty in generalizing about group tolerance to environmental conditions. The species variation among the five mayflies to only one physiological variable (DO) in this case is quite large.

As indicated earlier, some mayflies tolerate sedimented conditions under which they have adequate circulation of water over gills. Because some caddis flies (*Hydropsyche*) can undulate and thereby increase circulation, they are found in areas of intermediate DO. Dragonflies and damselflies (Odonata) are often found in silty conditions. They have modified rectal and caudal gills, respectively, which are also capable of increasing circulation and the efficiency of oxygen extraction.

The demand for DO has been found to vary greatly within the so-called intolerant groups, the stone flies, mayflies, and caddis flies (Figure 125). Thus, the concept of group tolerance is only very generally valid. On the other hand, oxygen consumption rates of these species, according to Olson and Rueger (1968), correspond well with their observed distribution in organically polluted streams. In order to use the

Figure 124. Respiration rate (O_2 consumption) versus O_2 concentration in four ephemeropterans. (Fox et al. 1935)

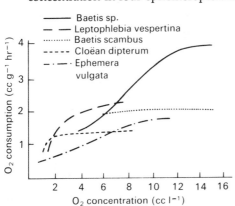

occurrence or absence of aquatic macroinvertebrates as an indication of oxygen availability, at least the genera, if not the species, level must be identified because of the variability among species. This point will be taken up later.

It is clear from experimental evidence in Macon (1974) that no single critical level of O_2 exists for macroinvertebrates in streams. The incipient limiting levels vary even within a family from very low to very high and vary even more with current velocity. For organisms adapted to high velocities (*Rhithrogena*), the lethal limit for DO increased greatly at low velocity; but for *Ephemerella*, which is morphologically capable of increased circulation at low DO and/or velocity (Ambühl 1959), the limit remained rather low and constant over a wide range of velocity.

Temperature

The range of tolerance for each critical stage in the organism's lifetime can be quite different for each species. Thus, an alteration in environmental temperature can result in changed community composition even if the ultimate lethal temperature in summer is not exceeded for any species. That is, a species can be eliminated from the community without mortality of adults. The replacement results from an altered ability to utilize food resources and thereby compete for living space. A general tolerance pattern for a hypothetical species is shown in Figure 126.

Although the range for survival of adults is quite large, the optimum range for growth usually occurs at a lower temperature. The reproductive stage and the early free-living stage in the development of the

Figure 125. Respiration rates of aquatic insects determined at 20 °C. Insects indicated in figure in descending order from left to right are *Tipula* (crane fly), *Caloteryx* (dragonfly), *Limnephilus* (caddis fly), *Pteronarcys* (stone fly), *Hetaerina* (dragonfly), *Paragnetina* (stone fly), *Macronemum* (caddis fly), *Ephemera* (mayfly), *Potamanthus* (mayfly), *Baetisca* (mayfly), *Leptophlebia* (mayfly). (Modified from Olson and Rueger 1968)

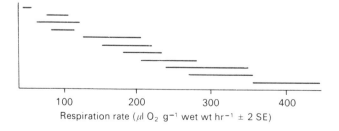

Respiration rate (μl O_2 g^{-1} wet wt hr^{-1} ± 2 SE)

young are critical. Gradual shifts in species dominance can occur as the *daily mean* temperature changes from normal. This has been shown to occur from quite small increases in temperature. Tolerance is apparently great for relatively wide diurnal fluctuations.

Although most of these principles have been demonstrated with fish, they probably also hold for most macroinvertebrates. In the determination of effect, the important factor associated with temperature change is time. A very high temperature can often be tolerated for a short time, whereas long-term survival can be guaranteed only at usually much lower temperature. Thus, when stating tolerance limits for any life stage of a given species the time of exposure must be included. Other factors that are almost as important include acclimation temperature, food availability, oxygen level, level of activity, age or developmental stage, and presence of inhibitory substances.

Criteria to determine temperature response and standards
Rate of rise. At an existing power plant the greatest threat to aquatic life is entrainment and passage with coolant water. In once-through-cooling, the entrained water can increase in temperature in a matter of minutes. Some evidence suggests that invertebrates can tolerate a relatively high ΔTt^{-1} (increment rise above ambient per time). Direct observations of the condition of entrained zooplankton show that rate of rise is not detrimental unless the final temperature reached equals or exceeds the lethal level. This is indicated by results summarized in Table 37. The lethal temperature for zooplankton seems to be the same

Figure 126. Generalized temperature tolerance graph for the important life stages of an invertebrate.

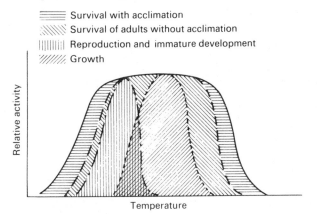

≡≡≡ Survival with acclimation
⑊⑊⑊ Survival of adults without acclimation
||||||| Reproduction and immature development
////// Growth

Relative activity

Temperature

whether subjected to abrupt or gradual temperature increase (Heinle 1969a; Welch 1969b). A similar result was found with *Mysis relicta* (Smith 1970), although the lethal temperature for that cold-water species was 16 °C.

Increment rise above ambient temperature. The total increment rise above ambient temperature (ΔT) can be important because reproduction in benthic animals in temperate areas is triggered largely by the achievement of an appropriate temperature rather than by a daily rate of rise. For the most part, temperature level is more important than photoperiod. Moreover, wide diurnal fluctuations in temperature can apparently be tolerated, whereas small changes in daily means cannot. As an example, a 7-yr study in a British pond showed that the dominant mayfly emerged when the temperature in the spring reached 10–11 °C, which occurred within a 2-wk period in April in spite of what seemed to be a wide variation in weather from year to year. The slow response of water bodies to climate change evidently makes water temperature a safer cue for aquatic animals than for terrestrial animals.

The seasonal changes (low temperature in winter and high in summer) are thus very important. The life cycles of aquatic animals are geared to seasonal temperature changes as well as to light. Crustaceans, such as the crayfish, must go through a low-temperature stage in which they do not molt but rather put their energy into reproductive cell development. Crayfish held in a cooling pond that was warm the year round did not stop molting and growing and did not reproduce until the winter temperature was reduced. The same response has been suggested with insects that suffered lower adult longevity and poorer emergence if held at constant but sublethal temperatures (Nebeker 1971a, 1971b). Also, emergence may occur as much as 5 mo early in

Table 37. *Zooplankton mortality compared to the maximum temperature and the ΔT*

	Percent mortality	Maximum	ΔT (°C)
Chalk Point Plant, Chesapeake Bay[a]	90	37	6.4
British plant on Thames[b]	0	24.4	7
Paradise Plant, Ky.[c]	100	35.5	9

[a] Heinle (1969b); [b] Markowski (1959); [c] Welch (1969b).

heated areas, which would be disastrous to adults in temperate climates (Nebeker 1971b).

Daily maximum. During the period of high temperature, the most damage results if the maximum reached exceeds the tolerable *level* regardless of the increment rise above ambient or the increment added to the normal river temperature. Heat addition to the Thames River resulted in a 12 °C rise above the ambient daily mean with a maximum of 28 °C. No change in the number of species present was observed, but the abundance of leeches, scuds (*Gammarus*), and midges decreased, whereas snails and clams increased (Mann 1965). The same rise above ambient in the Delaware River (12 °C), but with a maximum of 32-35 °C, resulted in an extensive reduction in the number of species as well as total abundance (Trembly 1960). Although the area repopulated in winter, the high temperature reached in summer was nevertheless damaging.

Community shift vs. temperature level

Evidence suggests that moderate toleration by aquatic freshwater invertebrates in general exists up to 30 °C. Figure 127 shows that the mode for the lethal temperatures of a large number of invertebrate species occurs around 35-40 °C using the available data. Allowing that a plotting of the preferred temperature (if available) would put the mode somewhat lower, say around 30-35 °C, means that a temperature increase to around 30 °C could result in an increase in species diversity

Figure 127. Distribution of freshwater invertebrate species tolerance to temperature. Dashed line shows approximate mode. (Bush et al. 1974 with permission of the American Chemical Society)

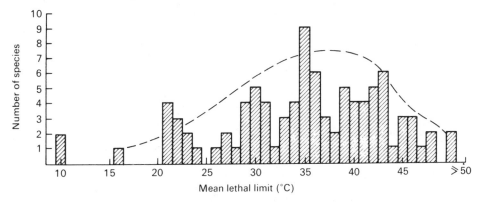

in some environments, but that increases to levels above 30 °C would probably result in decreases in diversity. This can be supported by the observations listed in Table 38.

The above limited sources of information suggest that to maintain normal diversity and abundance of most aquatic invertebrates the mean daily temperature should not exceed 30 °C. However, this does not mean that necessarily the same species would be present up to 30 °C.

For additional evidence to support the 30 °C level, a summary of temperature tolerances of 13 invertebrates and vertebrates in Chesapeake Bay is shown in Table 39. In this instance damage in terms of species reduction would presumably occur above 30 °C.

In the marine environment the maximum tolerable limit of mean daily temperature is no doubt lower than in fresh water because oceanic temperature is more stable and the ambient temperature does not reach the high levels that occur in fresh water. Studies are scarce in the marine environment. Adams (1969) cites results from a study of a

Table 38. *Observed tolerances of macroinvertebrates*

	Limit for normal diversity and abundance (°C)	Upper limit for highly tolerant forms (°C)
Delaware River	32	37
Chironomid midges	30–33	38–39
Caddis flies	28	35

Table 39. *Predicted percent of species in the suboptimal range and the percent lost with increases in temperature in Chesapeake Bay*

Temp. °C	Percent in suboptimal	Percent lost
26.7	0	0
29.4	8	0
32.2	61	8
35.0	16	69
37.8	15	85

Source: Mihursky (1969).

power plant on Morrow Bay, California, that showed predictable shifts from cold-water to warm-water invertebrate species with a temperature increase. A rise above ambient of 5.5 °C was observed at the surface as far away from the discharge as 150 m. At that point 54 species were present, 39% of which were warm-water types, which was considered normal for the area. At 90 m, of the 34 species observed 67% were warm-water types; near the heated-water discharge only 21 species were observed, 95% warm-water types.

Compared with some of the freshwater and estuarine examples cited, a 5.5 °C increase seems rather small. Nevertheless, the above examples suggest that marine communities may be very sensitive to temperature change.

The importance of such shifts in species as indicated for freshwater and marine environments depends upon the nuisance aspect of the invading species and the economic importance of the species affected. In general, however, a shift to warm-water species may be undesirable because warm-water forms grow fast and mature early but live short lives, whereas cold-water forms grow relatively slowly and mature late but live long lives. Nebeker (1971a) has shown for several insect species that even though feeding activity was high over a wide range (15 °C to 30 °C), emergence was increasingly less successful and the longevity of adults was shortened at temperatures much above 15 °C. Thus animals adapted to lower temperatures could be expected to reach larger size because they live longer. Although productivity would tend to be greater for warm-water forms, their quality for energy transfer would possibly be reduced and the overall stability of the system impaired.

Effect of food supply on macroinvertebrates

Benthic community composition depends upon the type and availability of food. If the type and/or availability of food changes, the community will respond even if pronounced changes in chemical characteristics do not occur. The following general food preferences indicate how changes in food supply could cause community shifts:

1. Detrital feeders can include the net-spinning caddis flies, aquatic sow bug, chironomids, clams, snails, some mayflies, and blackflies. These are largely collectors of fine particulate matter.
2. Grazers include most stone flies, some mayflies, case-building caddis flies, and snails.
3. Predators are represented by the dragonflies, leeches, a few stone flies, beetles, some midges, and a few caddis flies.

For example, a change in the food supply could affect stream invertebrate composition in the following way. In a stream of low productivity the amount of detritus resulting from primary producers is low, thus grazers utilizing periphyton on substrate surfaces are expected to be most abundant, with fewer collectors designed to utilize detritus accumulated on surfaces from autochthonous production. Now, if the amount of detritus is increased through increased autotrophic production or organic waste input, detrital feeders would be expected to flourish at the expense of grazers. That would be particularly true in the case of fine particulate inputs such as sewage. If poor water quality results from the increased production, such as low O_2, and thus limits predators, which ordinarily keep prey cropped low, the tolerant prey species could reach a large biomass in the predator's absence. Such an interaction has been noted with leeches and tubificid worms.

Effect of organic matter – natural inputs
Natural inputs of allochthonous organic matter can change community structure and actually result in species increases if the physical-chemical changes are not severe and diversity was previously limited by a scarcity of food. Terrestrial sources of organic matter are leaves, needles, and other forms of detritus. This was shown in results from a California mountain stream that receives particulate input (algal detritus) from a lake. At the point where particulate matter (seston) peaked from the lake enrichment, the number of species increased. Whereas 5 species were detected upstream from the lake, downstream the number increased to 11 and 14. Here the detrital-feeding blackflies (Simuliidae) were very abundant and midges (chiromids) increased in abundance. These are the "collectors" of detritus. Simulids removed an estimated 60% of the suspended plankton algae in 0.4 km of stream distance from the lake (Maciolek and Maciolek 1968).

Similar events are typical in the tail waters from reservoirs. Productive stream fisheries have resulted from the increased plankton detritus produced in the reservoir that subsequently stimulates the collector insect community downstream.

Effect of organic matter – wastes
The effect of man-made waste waters is usually more severe than the natural sources of allochthonous organic matter just described. What is the cause for the usual effect of organic waste? Rather than an

increase in species as well as biomass, the increase in biomass and productivity is accompanied by a reduction in species diversity.

The principal reason for the reduction is that the increasing severity of the physical and chemical factors eliminates the intolerant species. The remaining tolerant species flourish because of increased survival, a result of the absence of predators and a more favorable food supply, and, consequently, increased biomass results from increased food supply. But if food supply is too great for flow and O_2 resources, even biomass could decrease in septic zones. As in the example with natural organic matter, if the physical and chemical changes are not severe and the system is highly oligotrophic, increases in biomass as well as diversity could result from mild inputs of organic matter. Thus, the degree of community shift is a function of the type of waste, the waste load, the dilution volume or rate of river flow, and turbulence (reaeration potential). Sewage effluent produced these predictable effects in numerous studies. Results from one of the early but classic studies are shown in Figure 128. The associated BOD and DO results are shown in Figure 129.

Figure 128 shows the seasonal abundance of species upstream and downstream from a sewage outfall in Lytle Creek, Ohio. Note the relative difference in the total effect of the effluent when data from all seasons are considered compared with, for example, only data from August. The effect throughout the stream is much more severe in late

Figure 128. Distribution of species abundance in Lytle Creek, Ohio, upstream and downstream from a sewage outfall (arrow) at various seasons. (Gaufin and Tarzwell 1956)

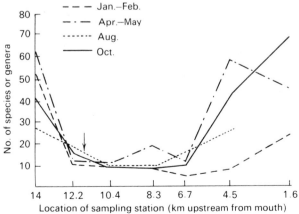

summer when flow is low. The higher BOD and lower minimum DO
in August that result from the low flow are shown in Figure 129. Also
note the extension of the zone of minimum species abundance during
the winter. This results from a reduced rate of waste decomposition
during the period of low temperature and high dilution.

Hawkes (in Klein 1962) has generalized the community response to

Figure 129. BOD and DO in Lytle Creek, Ohio, upstream and
downstream from sewage outfall (arrow) at various seasons. (Gaufin
and Tarzwell 1956)

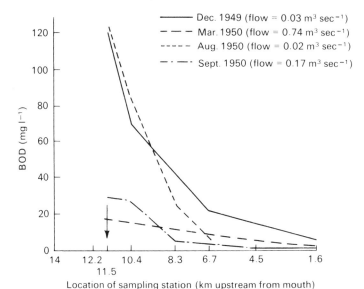

organic waste in the form of sewage in terms of the genera that are typical for stony, rapid streams as well as the depositing substrata type.

Decreases occur in intolerant species of stony rapids with the addition of organic waste according to

Rithrogena → *Ephemerella* (mayflies) → *Gammarus* (scud)

Decreased competition and increased food supply results in the following sequence of elimination:

Baetis (mayflies) → *Simulium* (blackfly) → *Hydropsyche* (caddis fly) → *Limnaea* (snail) → *Herpobdella* (leech)

Further increase in waste and substrate change favors dominance according to the following sequence:

Nais (worm) → *Asellus* (sow bug) → *Sialis* (alderfly) → *Chironomus* (midge) → *Tubifex* (worm)

Even in depositing substrata the end point is the same with *Sialis*, *Chironomus*, and *Tubifex*, but the initial community has sediment-tolerant organisms such as *Ephemera* and *Caenis* (mayflies).

Two oligochaete worms that dominate environments grossly polluted with organic waste are *Tubifex tubifex* and *Limnodrilus hoffmeisteri*. Although heavy loading of organic detritus and the accompanying low DO (and even total absence at times) contribute to the success of these species over all others in grossly polluted streams, Brinkhurst (1965) suggested that a principal reason for their dominance in large numbers can be largely ascribed to reproductive habits. In direct contrast to many of the "less tolerant" inhabitants of running water, these two species were found to breed at all times of the year. In fact, Brinkhurst suggested that the worms too are affected adversely by low DO and are able to maintain a large biomass in the presence of large food supplies in spite of frequent DO shortage because of frequent breeding.

The success of tubificids in utilizing and at the same time stabilizing organic waste loads can be illustrated by the potential sludge recycling rate of a high density population. At 100 000 m^{-2}, *Tubifex* can recycle the top 2-3 cm of sediment about three times per day. Each worm would, as a result, pass about 250 cm of fecal pellets.

Cairns and Dickson (1971a) have generalized the tolerance to organic waste into three principal groups. The intolerant group includes mayflies, stone flies, caddis flies, riffle beetles, and hellgrammites; the tolerant group includes sludge worms, certain midges, leeches, and certain snails; the moderately tolerant group includes most snails, sow bugs, scuds, blackflies, crane flies, fingernail clams, dragonflies, and some midges.

Although this grouping may appear rather crude, particularly when the overlap in tolerance among genera within families and species within genera is relatively great, conditions of gross pollution are separable from no or only moderate pollution with these groupings. In order to detect subtle changes in community response a more detailed approach is necessary.

Effects of toxic wastes

The effects of toxicity are much different from those of organic waste. The number of species decreases, as with organic waste, but biomass remains the same or increases slightly following the mortality of other forms that could represent or relieve a food source. The biomass can be maintained in the presence of only a very low level of toxicity. However, in most cases biomass will decrease as well as the number of species, and the effect of increased food supply is too minor to notice.

The degree of change in species number and biomass depends on the resulting concentration of toxicant in relation to the range in tolerance of the species present. Considering the foreign nature of most toxicants, particularly the synthetic ones, an increase in species number as predicted to occur from temperature and organic matter increase, in relatively cold and oligotrophic environments, respectively, would probably not be expected to occur with toxicant inputs.

A combination of both toxicants and organic wastes results in the predicted general community responses suggested by Keup (1966) and shown in Figure 130. In A of that figure the typical effect of a single organic waste is seen. Such a pronounced effect would necessarily result from a heavy loading, particularly when a sludge deposit occurs. Abundance increases, whereas variety decreases. In B the decrease in variety is similar to that which occurs with organic waste input, but the primarily toxic waste usually has little food value, and, consequently, the tolerant forms have no extra energy with which to increase their numbers. In C an initial toxicity exists, which "wears off" (complexed physically or biologically or diluted) before the full effect of the organic matter is observed. In D the level of the toxic substance(s) does not "wear off" and the combined effect is worse than with either alone.

Effect of organic waste via eutrophication in lakes

Oligochaetes increase in lakes relative to other organisms, particularly Chironomidae, as a result of cultural eutrophication. The worms also show greater abundance with depth, inversely to the trend of midges. This can be associated with oxygen depletion and sedimentation with depth, but the degree of species competition in causing the change is not known (Thut 1969). Note in Table 40 that Lake Washington compares closely with western Lake Erie and also the degree to which Lake Erie has been degraded from early surveys.

The density of oligochaetes can reach tens of thousands per square meter in nature. Carr and Hiltunen (1965) showed that the greatest densities in Lake Erie, which averaged 6000 m^{-2} in 1961, exist near the

Figure 130. Generalized response of macroinvertebrates to (A) organic waste, (B) low levels of toxic waste, (C) organic and toxic waste, (D) high levels of organic and toxic waste. Sludge deposit is indicated by cross-hatched area. (Keup 1966)

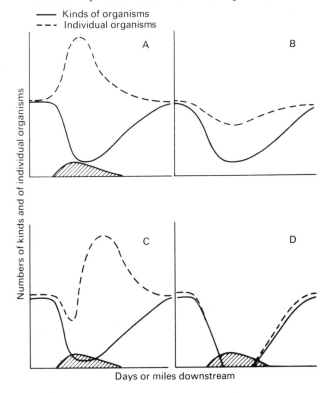

mouths of the Maumee, Raisin, and Detroit rivers. Densities in 1930 were less than 1000 m^{-2}.

The distribution with depth in Lake Washington is shown (Thut 1969) in Figure 131. Oligotrophic Cree Lake in Saskatchewan, Canada (Figure 132), does not show the proportional increase in oligochaetes with depth as occurs in Lake Washington. Thus, the proportional biomass change is an important indicator in lake eutrophication.

Species of chironomids have been associated with the trophic states of lakes. In the large lakes of Central Sweden, Wiederholm (1974) has suggested that four important species exist in that regard; the relationship is shown in Table 41. As can be noted, the levels of P and chl a suggest that these limits represent oligotrophy, mesotrophy, eutrophy, and possibly hypereutrophy, respectively.

Figure 131. The distribution of oligochaets and chironomids with depth in Lake Washington. (Thut 1969)

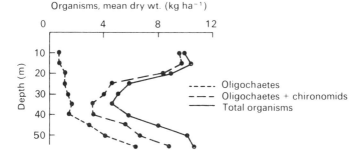

Table 40. *Relative composition of profundal benthic biomass in various lakes*

	Percent Oligochaetes	Percent Chironomids	Percent Sphaerids
Lake Washington	51	43	3
Lake Erie (1929–30)[a]	1	10	2
Lake Erie (1958)[b]	60	27	5
Cultus Lake (B.C.)	34	65	
Convict Lake (Calif.)	31	65	
Lake Constance (Calif.)	20	57	20
Lake Dorothy (Calif.)	23	69	3

[a] *Hexagenia* abundant; [b] *Hexagenia* absent.
Source: After Thut (1969) with permission of the Ecological Society of America.

Suspended sediment

The sediment carried by running water can have an adverse effect on bottom-dwelling organisms. As pointed out previously, the distribution of organism types and abundance is determined to a large extent by substrate type. Excessive erosion from abuse of the watershed – increased soil loss through overgrazing, deforestation, or various types of development or increased peak volume of runoff water from too much paved-over area, which results in within channel erosion and deposition – will create a depositing substrata. The greatest effect (change) on a stream benthos community will thus occur where increased deposition on an eroding substrata exists.

Suspended sediment is an insidious pollutant. The deterioration of

Figure 132. Distribution of chironomids and oligochaets in Cree Lake. (Data from Rawson 1959)

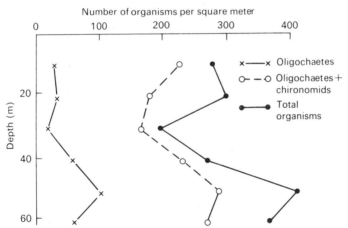

Table 41. *Relationship between four chironomid species and trophic state indicators*

	Hetero-trissocladius subpilosus	Micropsectra spp.	Chironomus anthracinus	Chironomus plumosus
Chl a μg l⁻¹	<3	3–10	10–20	>20
Total P μg l⁻¹	<15	15–30	30–60	>60
P load gm⁻² yr⁻¹	<0.5	0.5–1.0	1.0–2.0	>2.0

Source: After Wiederholm (1974).

a stream is gradual. No dramatic fish or invertebrate kill occurs. However, the community can be altered just as severely as if the waste were a toxicant. As such, the number of species as well as the abundance of organisms should decrease because there is little organic matter involved that could add to the food supply of deposit feeders. However, a moderate increase in suspended sediment to an eroding substrata could actually result in no appreciable change or in an increase in abundance, whereas the number of species could increase if the substrate were stable. Chutter (1969) found this in an African stream where in summer the number of species in a stony stream increased from 30 to 42 and then decreased to 22 as the substrate progressed from eroding to stable depositing to unstable depositing. The abundance for the three environmental types showed the same trend (Table 36).

An adverse effect on bottom fauna can result from rather modest increases in sediment concentrations. An effect has been observed in Bluewater Creek, Montana, an eroding substrate stream, as a result of irrigation return flows. The effect was noticed at an average level of around 100 mg l^{-1}. Although such a concentration of suspended sediment is not readily visible when viewed in a beaker of water, it considerably reduces light penetration and therefore visibility in a natural water course. There would be no toxicity result from such a concentration in a stream situation, yet the abundance of organisms can be depleted tenfold (Figure 133). The degree of effect from a particular concentration depends upon the retention in the stream substrata, which in turn depends upon river flow. Low-velocity streams will show greater insult from a given average sediment concentration than high velocity streams.

Recovery

An important question with regard to the management of water quality in streams is the extent and rate of recovery once waste input is curtailed. Relatively little information on this subject is available, largely because complete pre- and post-treatment monitoring has occurred on very few streams. This is essentially the same problem that plagues the restoration of lakes. Whereas considerable amounts of money are routinely spent on waste collection and treatment systems, very little is allocated for evaluation of treatment effectiveness in the stream. Yet the questions of whether and how much treatment is needed to recover streams are repeatedly asked in case after case.

Money for evaluation of receiving water recovery should be an integral part of each increment in waste-water treatment.

By and large, the recovery of stream systems is relatively rapid, because in a running-water system the contaminated water is moved downstream relatively quickly following curtailment of waste input. After one hydrologic cycle, which includes high flow that would carry away much of the contaminated sediment (or deposit new uncontaminated sediment), the physical environment should be largely recovered. All that is necessary then is repopulation, which requires reinoculation of the organism-depleted area and time for regrowth to pretreatment abundance. Because most of the forms have 1-yr life cycles, it is not surprising that 1- to 2-yr recoveries are common. Recovery could require a longer time in the lower reaches of rivers where sediment is not displaced readily or where most of the tributaries are depleted of fauna and reinoculation would thus be delayed.

Table 42 lists some examples of recovery in streams from various types of wastes. Most of these are on the order of 2 yr. With most of

Figure 133. The relationship between annual mean suspended sediment concentration and the abundance of benthic organisms at five stations in Bluewater Creek, Montana. Downstream (I-V) increase in sediment was caused by irrigation return flow. (Data from Peters 1960)

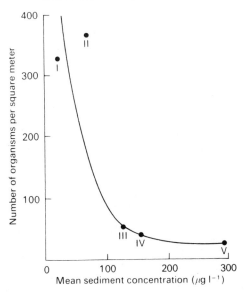

Table 42. Recovery of macroinvertebrate and fish communities affected by different types of wastes following waste controls

Stream ecosystem	Waste	State of recovery	Waste removed	Elapsed time
Clark Fork River, Montana[a]	Acid mine	Full recovery of trout and bottom fauna	100% of acid mine water after 6 months	2 yr
Maine, several streams[b]	DDT	Full recovery of bottom fauna	100% (discontinuous applications)	2-3 yr
Thames River estuary, England[c]	Sewage	Partial recovery; from 0 to 61 fish species, from 0 to 10% minimum O_2 saturation	60% sewage through secondary treatment	7 yr
Clinch River, Virginia[d]	Acid waste spill	Complete recovery of bottom invertebrate diversity, but some mollusks still absent	100%	2 yr
Clinch River, Virginia[d]	Alkaline fly-ash spill	Complete recovery of bottom invertebrate diversity	100%	6 mo
Blue Water Creek, Montana[e]	Silt	Marked improvement in trout/rough fish from 0.64 to 3.5 and 0.14 to 1.0 at two stations, respectively	52-44% of sediment load at two stations, respectively	2-3 yr
Plum Creek (Clink River)[f]	Fuel-oil spill	Complete recovery of bottom fauna abundance	100%	5 mo

[a] Averett (1961); [b] Dimond (1967); [c] Gameson et al. (1973); [d] Cairns et al. (1971b); [e] Marcuson (1970); [f] Hoehn et al. (1974).

these examples it is difficult to be anything but qualitative about changes in water quality and/or the bottom fauna composition. They nevertheless convey the impression that recovery is relatively "fast" if sources of reinoculation exist and the waste input is greatly reduced. The Clinch River fish kill in 1967, following a fly-ash pond leakage, resulted in a pH 12 water spill moving down river killing nearly everything in its path. Opinions, at the time, of a 10-yr recovery time were quite incorrect.

Assessment of water quality

The state of water quality can be effectively assessed by use of the benthic macrofauna. The advantages of this community were stressed earlier. An important question, however, is what methodology or technique should be used. Although the presence or absence of certain species has qualitative meaning to someone with experience in the relative tolerance of benthic organisms, such descriptive information is difficult to communicate or treat statistically. Thus, a knowledge of the relative tolerance of organisms present (and absent) is combined with a numerical expression of community structure, which may or may not treat the relative tolerances of one species versus another.

Indicator species and taxonomy

Indicator species are often employed to judge the relative quality of an environment, that is, when a certain species, an indicator species, is present a particular water quality condition is indicated. The absence of a particular species is also important, and although that is not implied in the approach, it is nonetheless often considered.

The principal drawbacks in this approach are taxonomic and the lack of definitive information on comparative tolerance to given types of waste. On the taxonomic problem Resh and Unzicker (1975) have stressed the importance of identification to species. They found that of 89 genera of macroinvertebrates that had known tolerances to organic waste, that is, intolerant (I), facultative (F), or tolerant (T), 65 have species in more than one general tolerance category (I/F, I/F/T, or F/T). Although the genera of invertebrates are relatively easy to determine in most cases, there apparently exists such an overlap in tolerance that a definitive analysis is not really possible unless identification is to the species level. With many groups this is difficult if not impossible because the state of taxonomy is relatively poor with North American fauna. For example, only about 20% of the North American caddis flies are described.

Equally important is the known tolerance of a particular species to the various types of wastes that can be discharged into streams. Only with organic wastes, such as sewage, is there any general understanding of tolerance among species.

This is not to say that nothing can be gained from a description of the organisms present, identified to some level short of species, and their relative abundances. As has already been shown, Hawkes (1962) described a typical succession of genera within English streams of differing substrata in response to organic waste. Certainly the presence of Tubifex in large numbers is quickly identified as gross organic pollution. However, the problem of detecting the effect of small incremental changes in waste input is probably difficult to solve by the qualitative indicator organism approach alone.

Numerical indexes

Formulas to express the "state of stream quality" or "degree of pollution" using the macroinvertebrates are numerous. Some formulas require a judgment as to the relative tolerance of the various taxa identified, usually tolerant, intolerant, of facultative, and the distribution among those groups is weighted (Beck 1955; Beak 1965). A major problem with these formulas is the lack of a general agreement or sound basis to categorize the tolerance of the taxa to the myriad of wastes that could be present. As a result, these formulas have not been generally adopted.

A more popular approach has been to quantify the community by simply enumerating the species present without a weighting for their relative tolerance. Although the number of taxa is very useful, it is also usually valuable to recognize the relative abundance of each taxon. For example, Gaufin (1958) found in the Mad River, Ohio, that although the snail Physa, the leech Macrobdella, and the worm Limnodrilus were present in clean-water reaches, they became very abundant in the reaches receiving organic waste.

The number of species and number of individuals can be included in a formula that denotes species richness (after Margalef 1958):

$$\frac{S - 1}{\log_2 N}$$

where S is the number of species and N is the number of individuals.

As one searches an environment and collects individuals, the number of species increases linearly as the logarithm of the number of individuals. At some point all of the species are discovered and the curve flattens out. This formula does not account for the relative abundance of each species, however.

Probably the most popular formula used to express the diversity of species is the one from information theory (Shannon and Weaver 1963):

$$H' = - \sum_{i=1}^{s} \frac{N_i}{N} \log_2 \frac{N_i}{N}$$

where s is the total number of species, N_i is the number of individuals in the ith species, and N is the total number of individuals. The unit for H' is bits per individual if $\log_2$ is used, which is customary in information theory because binary systems are of interest (Zand 1976). The maximum diversity H'_{max} from this formula is defined as $\log_2 S$.

Another similar formula, that describes diversity per individual is

$$H = \frac{1}{N} \log \frac{N!}{N_1! \, N_2! \ldots N_s!}$$

This formula is that of Brillouin (1962) and, according to Zand (1976) and Kaesler et al. (1978), it more accurately describes diversity when the number of individuals in each species is low. H' has been most popular and is the same as $\bar{d}$ used by Wilhm (1972).

Both of the above formulas treat species richness or abundance and the evenness or equability of those species, which are the ingredients of diversity. Thus, the environment with the more evenly distributed members, or individuals, for the numbers of species could be thought of as being better balanced. If an unpolluted environment receives a diversity of food materials and has a diverse bottom substrate, it should also have a more equitable distribution of the abundance among the species with differing requirements. Hulbert (1971) has criticized these formulations as having little biological meaning, because they essentially define the improbability of an event. He has suggested two formulas as being more biologically meaningful. These describe the probability of encounters and the relative importance of interspecific competition.

The formula defining $H'(\bar{d})$ will be illustrated here because it is the formula most often used for comparative study. It is emphasized, however, that many other formulas exist that may be more appropriate and it behooves the investigator to review these (see Hellawell 1977). Also, it will be illustrated that in many cases a simple account of the number of species may be more valid than a diversity index where species equability is treated as well as richness. Hulbert (1971) has indicated that richness can decrease while equability is actually increasing. This was illustrated by Cole (1973) where treated secondary effluent showed a decrease in the number of species but the H' values did not reflect the decrease. Apparently the substrate change caused by increased macrophytes in the stream resulted in increased habitat diversity and allowed for more evenness, which offset the decreased species richness. Often, however, one can expect to gain more sensitivity if evenness is considered along with richness rather than richness alone. Wilhm and Dorris (1968) provided a hypothetical example (Table 43) to show the influence of evenness together with richness.

Wilhm and Dorris (1968) have demonstrated that species diversity ($\bar{d}$ or H') is reliable in estimating the effects of wastes using benthic invertebrates as shown in Table 44. Near the waste outfall, $\bar{d}$ was typically in the neighborhood of 1. Intermediate values were found downstream with values farther down suggesting recovery.

Study of a variety of environments and waste types showed $\bar{d}$ to be a reliable index (Wilhm and Dorris 1968). The results consistently separated polluted and unpolluted conditions as follows:

| Clean water was represented by similar values for 22 widely varying environments | mean $\bar{d}$ = 3.30 range = 2.63–4.00 |

| Waste affected areas in 21 environments and widely varying waste types showed consistently low values | mean $\bar{d}$ = 0.95 range = 0.42–1.60 |

Table 43. *Diversity formula comparison for three hypothetical communities*

Communities	N_1	N_2	N_3	N_4	N_5	ΣN	S	$\dfrac{S-1}{\ln N}$	$\bar{d}$
A	20	20	20	20	20	100	5	0.87	2.32
B	40	30	15	10	5	100	5	0.87	1.67
C	96	1	1	1	1	100	5	0.87	0.12

Note: N = number of individuals in species 1–5; *S* = number of species.
Source: After Wilhm and Dorris (1968).

As a result of these observations Wilhm and Dorris have suggested
the following general guidelines for distinguishing the state of a
stream:

Heavy pollution $\bar{d} < 1.0$
Moderate pollution $\bar{d} = 1.0\text{-}3.0$
Clean water $\bar{d} > 3.0$

Although these are broad categories, and considerable degradation
may occur within the category "moderate pollution," it is nonetheless
useful to have some guideline. Whether such a guideline can be further
refined remains to be seen. As pointed out earlier, probably the greatest
danger in using a diversity index is a researcher's sole reliance on that
value to the exclusion of other approaches.

Figure 134. Diversity indexes for three taxonomic levels of
macroinvertebrates in response to an organic waste. (Egloff and
Brakel 1973)

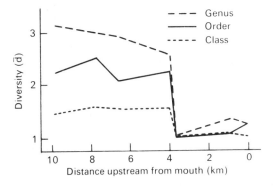

Table 44. *Comparison of $\bar{d}$ in response to various types of
wastes*

Waste	Above outfall	Outfall	Downstream	
Domestic & oil		0.84	1.59	3.44
Domestic & oil	3.75	0.94	2.43	3.80
Oil brines	3.36	1.58		3.84
Oil refinery		0.98	2.79	3.17
Total dissolved solids		0.55		3.01
Oil brines		1.49	2.50	
Oil brines		1.44	2.70	
Storm sewer		1.45	2.81	

Source: Wilhm and Dorris (1968).

Comparisons with general values would be particularly useful where preoperation surveys and operational (operation refers to the timing of industrial-waste discharge) control stations are not available and the general condition of the stream must be evaluated.

Whether identification is taken all the way to species for use with the diversity formula is optional. Egloff and Brakel (1973) have shown that one can use any level of taxonomy, genera, order, or even class, as long as identification is restricted to one level throughout the study. Although sensitivity decreases with the simplification of taxonomy, instream comparisons can nevertheless be made and waste effects analyzed (Figure 134). In this example from Plum Creek, Ohio, the decrease in $\bar{d}$ was associated with an increase in BOD from 3.5 to 9.0 mg l^{-1}. However, with anything besides the species level, $\bar{d}$ should not be applied to Wilhm and Dorris's general guidelines without qualification. That does not mean that each taxon needs a name but only that it should be morphologically distinguishable from all others.

Sampling problems

Although the abundance of benthic invertebrates is highly variable, diversity is much less so. Therefore, to use diversity to study waste effects in streams, select stations that are of similar substrate type, velocity, and so on and sample in a transect in either a transverse or longitudinal direction. Also bias the sample to high variety, or species richness, by selecting riffle areas that will increase the value of $\bar{d}$ and thereby the sensitivity. It is very difficult to obtain a representative sample for abundance and unnecessary for $\bar{d}$.

Diversity has been found to be largely independent of sample size (Wilhm 1972). The values for $\bar{d}$ were found to approach a maximum in the first 4 samples out of 10 in 13 different habitats. Thus, the mean diversity from 3 samples should provide a reasonable estimate of the maximum diversity in each section of stream, assuming that the number of individuals in each species is sufficiently great.

Study questions
1. Give one advantage for *detecting* the effects of waste waters biologically as opposed to chemically.
2. Compare the relative effects of sewage effluent in two hypothetical streams: A receiving effluent with secondary treatment (85% BOD removal) and B with no treatment. Indicate which

stream immediately downstream from the input shows the largest effect, the direction of effect (increase or decrease), and the reason with respect to each organism index.

   a. biomass of *Sphaerotilus* and other filamentous heterotrophic bacteria

   b. biomass of periphytic algae

   c. biomass of tubificid worms

   d. diversity of benthic invertebrates

   e. biomass of stone flies

3. The benthic invertebrate community of a stream showed the following changes downstream from a waste source over a period of three years: (1) diversity ($\bar{d}$) increased from 1.0 to 2.0 compared to a rather constant value of 3.0 upstream from the source; (2) biomass increased by threefold but was still less than the upstream station, and (3) periphytic algae increased to a slightly greater biomass than at the upstream station. What type of waste (organic, toxic, or sediment) caused the effect, had the quantity of waste changed over the years, and what is a reason for each of the three observed changes?

12

FISH

Fish are most frequently used as the so-called target organisms in setting standards for water quality. That is, the criteria for acceptable water quality are best defined with respect to fish. Specific standards exist for dissolved oxygen (DO) and temperature. These will be evaluated here. The standards for toxicants are usually undefined, however, and the limits are determined with bioassays, either in situ or in vitro with the relevant water and species.

Dissolved oxygen criteria

Suggested standards for DO have been given by the National Committee on Water Quality (*Water Quality Criteria* 1968; 1973) to protect communities and populations of fish and aquatic life against mortalities as well as prevent adverse effects on eggs, larvae, and population growth. A distinction has been made between cold- and warm-water environments. Concentrations should stay above 5 mg l^{-1} nearly all the time, assuming periods of much higher concentrations. For short periods, concentrations of 4–5 mg l^{-1} should be tolerated. In cold-water environments concentrations should stay above 7 mg l^{-1} for successful spawning and egg and larval development. Greater than 6 mg l^{-1} should exist for growth, and greater than 5–6 mg l^{-1} for survival over short periods of time.

Minimum levels for adult survival may be surprisingly low for even fish of low tolerance to DO depletion. For example, 1–3 mg l^{-1} is sufficient for survival for a short time even with cold-water species such as salmon. These same fish could feed and reproduce at 3 mg l^{-1} (Doudoroff and Shumway 1967). However, the level required to sustain the normal production of highly desired species, such as salmon and trout, is much higher.

Early field evidence by Ellis (1937) showed that DO concentrations greater than 5 mg l^{-1} were associated with populations of fish of good abundance and species diversity. That was largely the basis for the early setting of the 5 mg l^{-1} standard, which has often been interpreted as a "requirement" for fish. However, the limit of 5-6 mg l^{-1} that has come to be accepted in standards is probably a compromise to accommodate either warm- or cold-water fish. Requirements for some stages of a fish's life history are much higher. To ensure maximum production of many desired species of fish, the DO content at air saturation (9.2 mg l^{-1} at 20 °C) should be maintained for the critical stages of life (Doudoroff and Shumway 1967; *Water Quality Criteria* 1973). However, many nongame-fish species, such as goldfish (*Carassius*), have their normal activity limited at much lower concentrations of DO than do the salmonids (Macon 1974).

Before proceeding to a discussion of effects of low DO on the various life stages and activities of fish, it would be well to make a point about the units with which to express DO. Although fish obtain oxygen from the water according to the partial pressure difference between the surrounding water and the blood circulating through the gills, the desired unit is concentration rather than partial pressure or percent saturation (Doudoroff and Shumway 1967), because the metabolic demand for oxygen by fish increases as the temperature increases. At the same time the actual concentration at 100% saturation decreases. That is, the amount of DO necessary to maintain a constant partial pressure (100%) decreases as temperature increases.

That fish actually need more DO (concentration) as temperature increases to carry out normal activity, and as a result much more than a constant partial pressure, is shown in Figure 135, which was interpolated from data by Graham (1949). There it can be seen that the standard metabolic rate of brook trout increases dramatically as temperature increases to near the lethal limit. At the same time, the minimum DO concentration to sustain that metabolic rate increases from about 2 to 4 mg l^{-1}. The minimum saturation level, of course, increases to an even greater extent.

Embryonic and larval development

The growth of salmonid (coho) embryos and the size of emerging fry, which would normally occur in the gravel of streams, are limited by the supply of DO as shown by Shumway et al. (1964) in Figure 136. The dotted lines indicate embryo growth at 80% and 67% of the maximum.

Note that there is an interaction between DO and water velocity on the size of the emerging fry. A given level of growth can be sustained by either a velocity increase with declining DO or a DO increase with declining velocity. At a constant high or low velocity, growth shows a progressive, although gradual, decrease below air-saturation values to a level of 5 mg l^{-1} after which the decline in growth is more rapid.

Later work (Carlson and Siefert 1974; Carlson et al. 1974; Siefert and Spoor 1973) has shown a similar effect of DO on embryonic development to the larval stage. For most species studied, which included a range from cold-water to warm-water types, the time to hatching increased, growth decreased, and survival decreased as DO was reduced. The greatest decrease occurred at about 50% saturation, which was usually 5-6 mg l^{-1} at the test temperatures used. With species such as the white sucker, detrimental effects were not observed unless the DO

Figure 135. Standard metabolism of *Salvelinus fontinalis* in relation to minimum DO needed in concentration and percent saturation. (Modified from Graham 1949)

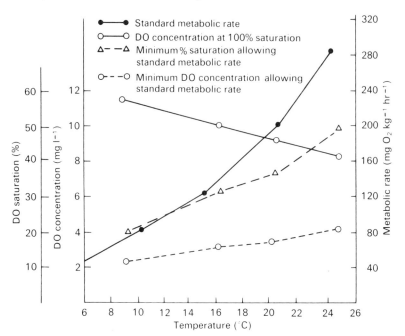

level was below 50% saturation. Data on survival for these species are shown in Figure 137. Although it is clear that survival dropped off most precipitously at concentrations below 5 mg l⁻¹ for all species, the three species with the poorer survival at high concentrations (salmon, trout, and catfish) also tended to show a reduction in survival as DO was reduced below the highest level. Furthermore, delayed hatching, delayed feeding, and/or slower growth were also shown with these species at levels between 100% and 50% saturation. These results, therefore, do agree with those of Shumway et al. (1964); that is, detrimental effects occur with cold- and warm-water species alike when any reduction occurs in DO to levels below 100% saturation even though the greatest damage occurs at concentrations below 5 mg l⁻¹.

Swimming performance

The swimming performance of coho salmon also begins to decline progressively below 100% saturation or at about 9–10 mg l⁻¹ DO (Figure 138). Again there is no clear threshold of effect, although

Figure 136. Development of coho salmon embryos at different concentrations of DO and at different velocity levels. Dashed lines indicate DO and velocity combinations producing 20 and 33% reduction in growth. (Shumway et al. 1964)

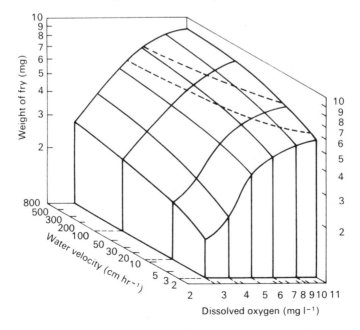

the decrease in performance, as with embryo development, is greatest at DO content below 5 mg l⁻¹. The points represent velocities at which the salmon failed to maintain themselves in a current (see Davis et al. 1963 for testing details). Therefore, at DO content below 9–10 mg l⁻¹ the ability of migratory fish to traverse velocity obstacles or for any fish to avoid predators would be reduced (Davis et al. 1963).

Food consumption and growth
    The amount of food consumed, as well as growth, is also progressively impaired below saturation or concentrations of 9–10 mg l⁻¹ DO. This was found to be true for juvenile largemouth bass (warmwater fish) as well as salmon (Warren 1971). Figure 139 shows the relationship of food-consumption rate with DO concentration for bass (Stewart et al. 1967).

Fish probably respond to the minimum daily DO rather than the daily average as shown in Figure 140. For fish held at fluctuating diurnal concentrations, the resulting growth was not much different

Figure 137. Effect of DO concentration on survival of fish larvae from the embryo stages for four species. Time periods and temperature were as follows: channel catfish, 19 days at 25 °C; lake trout, 131 days at 7 °C; coho salmon, 119 days at 7–10 °C; largemouth bass, 20 days at 20 °C; and white sucker, 22 days at 18 °C. (Data from Siefert and Spoor 1973; Carlson and Siefert 1974; Carlson et al. 1974)

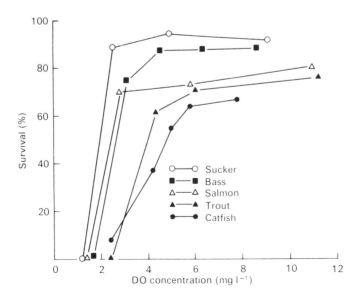

Figure 138. Relationships between DO concentration and swimming velocity in coho salmon at 20, 15, and 10 °C. Each point respresents velocity at which first underling failed to maintain orientation. (Davis et al. 1963)

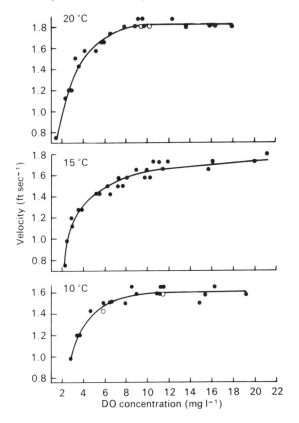

Figure 139. Food consumption in largemouth bass related to DO concentration. Each curve represents a separate experiment. (Modification from Stewart et al. 1967)

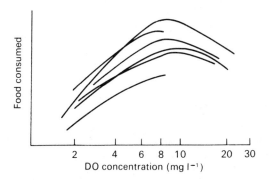

from the growth of fish (coho salmon) held at a constant DO near the minimum in the case of a diurnally fluctuating DO. At unrestricted food ration, it is clear that salmon require a relatively high DO concentration for maximum growth. This is largely a function of the more food consumed, the more oxygen necessary for full utilization.

To be sure, the requirements for DO vary among different species and seasons. However, the previously cited information indicates that near-saturated concentrations of DO are required during reproductive periods and periods of high food availability, consumption, and growth if maximum production of cold-water as well as warm-water fish is to be realized. Much less than saturation can be tolerated during periods of reduced feeding, and growth is nearly independent of DO if not feeding. The minimum limits of DO at 5–6 mg l⁻¹, often cited in standards, carry an assumption that normal variations result in concentrations being much higher than 5 mg l⁻¹ most of the time. Doudoroff and Shumway (1967) have suggested that biologists should refrain from giving a value that is supposed to "sustain maximum production of natural populations" unless it is *air saturation* (9–10 mg l⁻¹ for cold-water fish). Of course, it is recognized that such high values are not needed all the time. In order to develop more specific standards, more detailed seasonal requirements for various species would need to be determined. Table 45 shows a working guideline based on the hypothesis of progressive detrimental effect below saturation (*Water*

Figure 140. The growth rate of juvenile coho salmon fed restricted and unrestricted rations at constant and diurnally fluctuated oxygen concentrations. (Doudoroff and Shumway 1967)

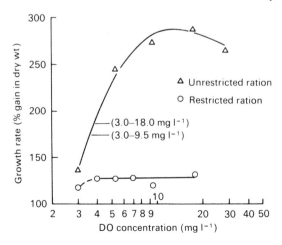

Quality Criteria 1973). Minimum levels are suggested to provide high, moderate, and low levels of protection.

These suggested levels are based on the concept that some damage will result from any reduction in the average daily minimum DO below air saturation. From the appearance of many of the relationships (Figures 136–140), damage may not be as linearly related to DO reduction as this model (Table 45) would suggest. Nevertheless, this is an approach to a more reasonable guideline than simply the traditional 5 mg l^{-1}.

DO and sediment

The effects of reduced DO supply on developing trout eggs through sedimentation from stream-bank erosion at peak flow and sediment in irrigation return flows were evaluated in Bluewater Creek, Montana (Peters 1967). This effect is similar to that caused by urban runoff patterns from a high proportion of impervious cover although part of that effect may be caused by physical damage to eggs in unstable substrate. More than 80% mortality of trout eggs was observed at stations where water carried a mean sediment content of 200 mg l^{-1}. This is similar to the effect of sediment that enters salmonid streams as a result of deforestation and the accompanying road-building activities. With the filling in of the interstices within the bottom substrate with fine sediment, the intragravel water velocities are reduced. As those velocities decrease, the supply rate of DO to developing embryos is reduced as suggested in Figure 136. At the same time the DO of the

Table 45. *Example of recommended minimum concentrations of DO*

Estimated natural seasonal minimum DO	Recommended minimum DO for selected levels of fish protection			
	Nearly maximum	High	Moderate	Low
5	5	4.7	4.2	4.0
6	6	5.6	4.8	4.0
7	7	6.4	5.3	4.0
8	8	7.1	5.8	4.3
9	9	7.7	6.2	4.5
10	10	8.2	6.5	4.6
12	12	8.9	6.8	4.8
14	14	9.3	6.8	4.9

Source: Water Quality Criteria (1973).

intragravel water is apt to decrease because of the increased retention time of intragravel water (or slower replenishment of oxygen-saturated surface water) and the possible increased oxygen demand of the sediments from settled organic matter.

Clearcutting watersheds of small streams has resulted in DO decreases of 40% in the intragravel water that persisted at least three years after logging (Hall and Lanz 1969; Ringler and Hall 1975). Pre- and post-logging (three years after) intragravel DO averaged 10.5 and 6.2 mg l^{-1}, respectively, during the winter–spring period. That resulted in a 73% reduction in the resident cutthroat trout population. Immediately after logging, intragravel DO dropped to 1.3 mg l^{-1} and stream temperature reached 24 °C in summer. In a partially clearcut watershed (25%), where buffer strips were left along the stream in affected areas and trees were not felled in the stream, DO decreased only about 10% and no damage to fish was observed. Emergence of young coho salmon was apparently unaffected by the physical changes.

In Bluewater Creek the banks were stabilized with vegetation, and irrigation return-flow ditches were lined to cut down on sediment input. The results after 2 yr of stream recovery are shown in Table 46 in terms of the changing trout/rough fish ratio (an increase of sixfold).

This illustrates the significance of stream-bank stabilization, vegetation protection, and flow control that could be employed to preserve the quality of urban streams. Also, streams can show some degree of recovery from sedimentation. Of course, the most pronounced recovery would come in those streams with a significant discharge velocity and only a moderate natural bed load of sediment. Those with a rather natural depositing substrata would probably show less improvement.

Table 46. *Sediment load reduction and fish population improvement in Bluewater Creek, Montana; stations numbered in a downstream direction*

Station	Sediment reduction (T day^{-1})	Percent reduction	Pretreatment trout/rough fish	Post-treatment trout/rough fish
2	1.9	32		
3	14.0	52	39/61	78/22
4	10.5	44	12/88	51/49

Source: After Marcuson (1970).

DO, eutrophication, and fish

The principal effect of eutrophication on fish is one of DO depletion. An increase in plankton production to within the meso-trophic state may well have a beneficial effect on the production of desirable species of fish, that is, cold- and warm-water sport fish alike. However, once the state of eutrophy is reached, and particularly if the lake stratifies in summer, there is a strong probability that DO in the hypolimnion may reach a critical level for survival toward summer's end.

As eutrophication increases, the minimum DO reached will continue to decrease and the minimum DO will occur earlier. That situation will be increasingly detrimental to particularly cold-water fish but also to warm-water fish, because it is known that activity and growth decrease progressively with decreasing DO. Both types will tend to evacuate the epilimnion when temperatures exceed their preferred level. If adequate DO exists in the hypolimnion, it can be a healthy refuge during the warm summer period. If there is inadequate DO, the fish will be subjected to either stressful DO in the hypolimnion or, if excluded from the cooler but oxygenless hypolimnion, to stressful temperature in the epilimnion.

The higher the nutrient loading to the system, the greater the DO deficit, and depending upon the hypolimnetic depth, the lower the DO concentration and, consequently, the more inhospitable the entire lake will be to fish. Even shallow lakes that do not stratify will suffer from low DO in winter if they become ice covered. The problem of "winter kill" in shallow lakes is a long-standing one in temperate areas (Halsey 1968). Also, following algal bloom die-off in summer low DO can cause fish mortality (Ayles et al. 1976).

The long-term effect of eutrophication will be one of changed species composition, a result largely of the changed DO status. However, in-creased food supply in the form of more detritus, which tends to lead to smaller zooplankton and a less diverse worm/midge-dominated bottom fauna, would tend to favor detritus/bottom-feeding fish such as suckers and carp. Haines (1973) has shown that the growth of carp was much greater (nearly four times) and smallmouth bass much less (factor of about 2 to 6) in fertilized than in unfertilized experimental ponds. The fertilized ponds received phosphorus at a rather high rate of about 2 g m^{-2} per year. DO was probably the principal factor detrimental to the bass because the diurnal range was from 18 to less than 2 mg l^{-1} in the fertilized ponds, but seldom exceeded a range of 3.5 mg l⁻¹ in the unfertilized ponds.

Regardless of which factor is more important, DO or changes in the food supply, the results are similar. In Lake Erie, for example, the populations of cisco, whitefish, walleye, sauger, and blue pike drastically declined over the 40-yr period that loading of nutrients to the lake was increasing, as indicated by increases in major ions (Beeton 1965). (The principal cause for their decline was a failure to reproduce.) The total fish catch, however, did not decline; rather, the catch of the desirable species was replaced with such species as carp, perch, buffalo, drum, and smelt (Table 47). Although other factors, such as fishing pressure and sea lamprey predation, were involved in changes in Great Lakes fish production, Beeton and Edmondson (1972) suggested that the changes in Lake Erie, and in particular the cisco, were closely tied to the progressive eastward movement of polluted conditions.

This same pattern can be expected to occur in most lakes undergoing eutrophication, depending on the oxygen resources, which are a function of depth. Eutrophication is considered to be the principal problem that has limited salmonid fisheries in European subalpine lakes (Nüman 1972).

The general pattern of change is shown in Figure 141. Increased abundance and productivity of "desirable" fish species should occur in the mesotrophic stage. As eutrophy is approached, those species tend to diminish and are replaced by a dramatic increase in production of the more DO- and/or temperature-tolerant "undesirable" species.

Table 47. *Comparative annual catch of commercial fish species in metric tons over a 40-yr period in Lake Erie*

Species	Early years	Recent years
Cisco	9 000 (pre-1925)	3.2 (1962)
Sauger	500 (pre-1946)	0.45–1.8
Blue pike	6 800	0.45 (1962)
Whitefish	1 000	6.0 (1962)
Walleye	7 000 (1956)	450
Total	24 300	
Drum, carp, perch, and smelt		~22 240 (Increased catch)
Total		22 700

Source: Data from Beeton (1965).

There is little doubt that a eutrophic state is definitely detrimental to the production of the more DO/temperature-sensitive species. The question is how much enrichment is detrimental? Are the criteria previously described for trophic states in relation to recreation also pertinent to fish propagation? Apparently the rate of phosphorus loading that is apt to cause a eutrophic state from the standpoint of algal biomass and transparency is also similar to the loading that will cause an ODR (oxygen deficit rate) that is representative of eutrophy (see Chapter 9). Therefore, the oxygen resources of a lake begin to be strained from the standpoint of the fishery at a P loading rate similar to that at which the recreational opportunities are impaired (Welch and Perkins in press). The meaning of eutrophy, therefore, is similar to the fishery and to recreation. Dillon and Rigler (1974b) have placed very conservative guidelines for salmonid fisheries in Ontario lakes with respect to increased P loading and plankton algae.

Fish and temperature

The thermal tolerance of fish, and other aquatic animals as well, can be illustrated by Figure 142. A very large zone of tolerance can be defined within which all life-history stages of a species can

Figure 141. Suggested changes in various characteristics of lakes with eutrophication. (Modified from *Lake Erie Report* 1968)

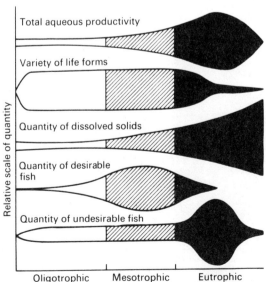

Total aqueous productivity

Variety of life forms

Quantity of dissolved solids

Quantity of desirable fish

Quantity of undesirable fish

Relative scale of quantity

Oligotrophic   Mesotrophic   Eutrophic

proceed normally. The lower and upper tolerance limits, often referred to as the upper and lower incipient lethal levels, increase with increases in the acclimation temperature within the range of tolerance. At high temperatures, there is a zone of resistance that is rather narrow between the incipient lethal level (long term) and the lethal short-term limit. The increase in the lethal level for roach was found to be 1 °C for each 3 °C increase in acclimation temperature over the range of tolerance (Cocking 1959).

Lethal temperature
Although the lethal limit is not too useful as a maximum limit to control the effects of heated water, the values are determined with a high level of precision and do indicate the progression of species tolerance to a wide temperature range. Figure 143 shows the distribution of fish families to lethal temperatures over the range of tolerance. Several points are apparent. Smelts (Osmeridae), salmonids, and whitefish (Corigonidae) are the most intolerant of high temperature, whereas catfish (Ichtaluridae) and some minnows (Cyprinidae) are the most

Figure 142. Zones of tolerance and resistance of fish as affected by changing acclimation temperature. (Modified from Brett 1960)

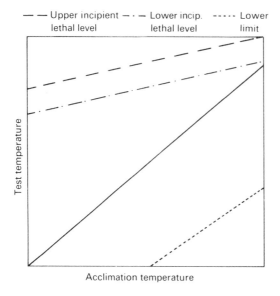

— — Upper incipient — · — Lower incip.   - - - - - Lower
lethal level   lethal level   limit

Test temperature

Acclimation temperature

tolerant. The minnows are, however, quite variable in their tolerance. There are so many species of minnows it is not surprising that the members of that family are found in (and tolerate) a wide variety of thermal regimes.

The lethal limits are largely determined as a $TL_m{}^{96}$-median tolerance limit, over time. Thus, they designate a constant exposure temperature, usually interpolated, at which 50% of the test fish survive for 96 hours. If the temperature is fluctuated daily, slightly higher maximums can be tolerated for short periods if temperature is lowered below that level for longer periods. This has been demonstrated with juvenile salmon exposed to raised and lowered temperature (Figure 144). In this case, example C produced the greatest mortality, probably because the mean temperature of exposure was greatest over the exposure period even though in each case the fish were exposed to greater than the lethal temperature for a short period.

Figure 143. Families of fishes including number of species and lethal temperatures. (Compiled by Welch and Wojtalik 1968)

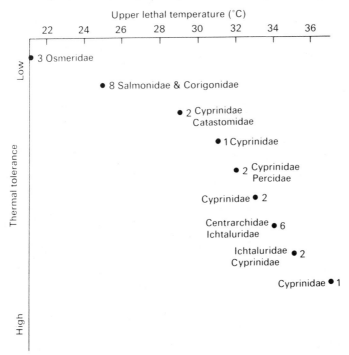

Although fish can resist temperatures above the 96-hour lethal limit for short periods, such thermal shocks have been shown to be detrimental to salmon even if not thermally lethal (Coutant 1969). At temperatures from 28 to 29 °C salmon showed measurably shorter times to "equilibrium loss" than to death and this equilibrium loss resulted in increased predation on the juvenile salmon.

Limits for normal activity and growth

Obviously, thermal standards or limits should not be set solely on lethality or "equilibrium loss" in environments where fish are expected to grow and reproduce. The "scope for activity" is the temperature range at which fish can feed, swim, and avoid predators at an optimum rate. The scope for activity (as determined by respiration rate) occurs at a temperature range much less than the lethal limit (25 °C) in brook trout, whereas for bullhead catfish the maximum scope occurs near the lethal limit, which is 37 °C. This example illustrates a difference between cold- and warm-water fish (Figure 145). The position of the maximum scope for activity for brown trout is similar to that for bullhead, but that trout's lethal limit is near 25 °C, similar to brook trout. Brown trout are generally considered more tolerant of higher sublethal temperatures than brook trout (Brett 1956). This may be an important reason why brown trout often tend to dominate in apparently marginal waters.

Warren (1971) has discussed how the "scope for growth" for salmon should be maximum at temperatures much less than the lethal limit, but that growth is also a function of food availability. Growth will decrease irrespective of temperature when food is restricted; however, the temperature range for the optimum scope will become narrower when food is restricted. Averett (1969) has supported this hypothesis with juvenile coho salmon, showing that the optimum temperature or maximum scope for growth is in the range 14–17 °C (Figure 146). At unrestricted ration, the amount available for growth, after the necessary metabolic demands were met, decreased above 17 °C, which is 7 °C

Figure 144. Effect of exposure to three kinds of temperature cycles on mortality in juvenile salmon. (Data from Templeton et al. 1969)

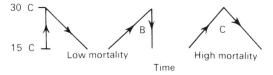

below the lethal limit for coho. The normal metabolic demands or losses were cost of food handling, standard metabolism, and excretion.

Although temperatures for maximum scope for growth and activity would be desirable as criteria for all important and sensitive species, such data are difficult to obtain. A more easily obtained index is available, however. "Preferred temperature," or the temperature willingly selected by a species under experimental conditions, is the temperature that most approximates the temperature that allows the maximum scope for growth and activity. Table 48 shows lethal and preferred temperatures for cold- and warm-water species. Note that the preferred temperature for brook trout is 14–16 °C, very close to the temperature for maximum scope for growth of coho salmon (14–17 °C). These data are plotted in Figure 147 and show that the preferred temperature becomes closer to the lethal limit as the thermal tolerance increases from cold- to warm-water species.

That fish select a preferred temperature in nature was demonstrated in a study of fish distribution in the Wabash River, Indiana, in response to a heated-water discharge (Gammon 1969). The effluent was on the average 8 °C higher than the river temperature. The mixed temperature within 1.2 km downstream was 2–3 °C above the river temperature. Good agreement was observed between temperature preference determined in the laboratory and the distribution of species in thermal zones in the river (Table 49). Although a clear segregation of

Figure 145. Relation of temperature to metabolic rate and scope for activity in brook trout and bullhead catfish. For brook trout, lethal temperature is 25 °C and peak scope for activity is at 19 °C. For bullhead, lethal and peak scope temperatures are at 37 °C. For these species, scope for activity is difference between active and standard metabolic rate. Note that scope for activity is much greater for bullhead than trout near the lethal limit. (Brett 1956)

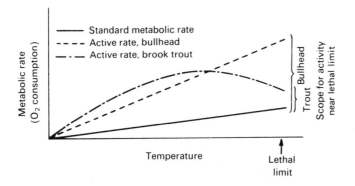

species was observable during the warm season in relation to mean temperature, during winter all species were attracted to the effluent area because the resulting temperature was closer to their preferred temperature than the normal river temperature.

The abundance of all species in the river in relation to the mean daily maximum temperature is shown in Figure 148. The number of species was substantially reduced in the zone of maximum temperature increase. Although the temperature considerably downstream of the

Figure 146. Partitioning of food energy consumed in coho salmon at varying temperatures. Crosshatched area shows energy as growth, diagonal line indicates total energy consumed. (Modified from Averett 1969)

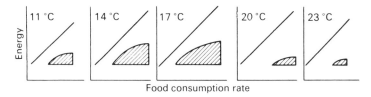

Food consumption rate

Table 48. *Final temperature preferred and some upper lethal temperatures for a range in species tolerance*

Species	Final preferenda (°C)	Upper lethal (°C)
Lake trout	12	
Lake whitefish	12.7	
Rainbow trout	13.6	
Brook trout	14–16	25
Brown trout	12.4–17.6	
Yellow perch	21	
Burbot	21.2	
Muskellunge	24	
Yellow perch	24.2	32
Grass pickerel	26.6	
Smallmouth bass	29	
Goldfish	28.1	34
Pumpkinseed	31.5	
Carp	32	
Largemouth bass	30–32	34
Bluegill	32.3	34

Source: Welch and Wojtalik (1968).

maximum heated zone was higher than the normal river, more species were actually found there. Fish do not necessarily exist only in areas of their preferred temperature. Although all species existed in the river before heat addition, there were actually more that preferred a slightly elevated temperature than the normal. An important management point, however, could be that the most desirable species is one of the most sensitive. In this case, smallmouth bass, a highly prized game species, avoided all temperature increases.

Figure 147. Preferred and lethal temperatures of five genera of fishes. (Compiled by Welch and Wojtalik 1968)

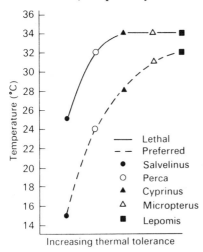

Table 49. *Distribution of representative species of fish in the Wabash River among several definable thermal zones*

Avoided all increases	Thermal gradient in the river		
	Normal to 27°C	32°C	35°C
Golden redhorse	Sauger	Gizzard shad	All fish left
Short head redhorse	White bass	Carp	the area,
Smallmouth bass	Northern river	Buffalo fish	no
Spotted bass	carpsucker	Longnose gar	mortality
		Shortnose gar	
		Channel cat	
		Flathead cat	

Source: Data from Gammon (1969).

Based largely on preferred and lethal temperatures, the prediction of fish community changes in the Columbia River (cold water) and the Tennessee River (warm water) were hypothesized with increasing (mean daily) maximum temperature (Bush et al. 1974). If the percent of species lost is plotted versus temperature for both rivers and compared with the normal maximum temperatures (Figure 149), it is again indicated that fish in warm-water environments are living closer to their lethal limit than most species in cold water. In reality then, a greater

Figure 148. Relationship of fish species number to temperature increase in the Wabash River, Indiana, caused by heated water from a power plant. (After Gammon 1969)

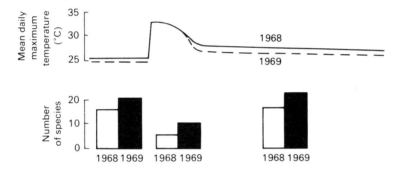

Figure 149. Predicted species loss rate from a cold-water system (Columbia River, maximum mean daily temperature 21 °C) and a warm-water system (Tennessee River, maximum mean daily temperature 30 °C) with increasing temperature. (Bush et al. 1974)

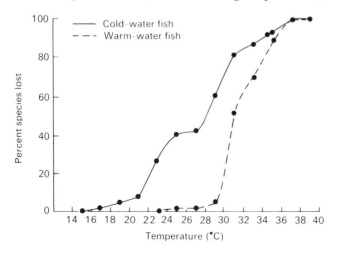

"assimilative capacity" for heat does not occur in warm water just because those species are more tolerant. On the other hand, there is not that much assimilative capacity available in cold water because the maximum scope for growth and activity tends to be much less than the lethal limit in contrast to warm-water species. Thus, although the margin available for temperature increase in cold water may be only a few degrees, because of a few important species like salmon and trout, actually more species would be lost by an equivalent increase in warm water.

Thermal limits for reproduction have been determined rather precisely for salmon but not for warm-water species. The results given in Table 50 for salmon were summarized by Brett (1956). The following limits seem warranted as limits for the respective warm-water species even though experimental evidence is largely lacking:

Largemouth bass   19 °C
Smallmouth bass   16.5 °C
Walleye pike   7-10 °C

Brett (1960) early on summarized general life-stage and seasonal thermal requirements for pacific salmon (Figure 150). Note that the most restricted range is during egg development and that a gradual increase is recommended in spring for normal growth to result. The temperature for optimum growth is suggested not to change, but the lethal temperature is seen to increase in spring and summer, probably as a result of acclimation to rising temperature and a longer photoperiod.

Temperature standards

The standards for temperature recommended by a committee on water quality (*Water Quality Criteria* 1968) generally separate cold-

Table 50. *Thermal limits for salmon reproduction*

Species	Limits for successful hatching and survival of young (°C)
Sockeye	4.4-5.8 to 12.8-14.2
Chinook	5.6-9.4 to 14.4
All salmon	5.8 to 12.8

Source: Brett (1956).

and warm-water environments. Warm water is defined by a >25 °C maximum mean daily temperature, so defined here because 25 °C is the upper lethal limit for salmonids. The recommendations are that the rise above ambient (normal, before heat addition) should not exceed 2.8 °C in streams and 1.7 °C in lakes. The maximum temperature at any time should not exceed 32–34 °C, depending on area. The rate of rise should not exceed 3 ° per hour (not too commonly applied). Cold water is conversely defined by a < 25 °C maximum mean daily temperature. The rise above ambient should not exceed 2.8 °C in streams and 3 ° in lakes. The maximum temperature at any time should not exceed 20–21 °C.

Comments on standards

The recommended maximum temperatures of 34 °C and 21 °C are the limits of tolerance of warm- and cold-water species as noted earlier (Figure 148). If that temperature (34 °C) persisted for as long as a day in warm water, mortality could result or, more likely, many species are apt to leave the area.

The maximum rise above ambient, or ΔT, is the second most critical factor and must be set from consideration of the following:

1. If the maximum temperature at any time is not allowed to exceed the recommended limit, the length of exposure to that summer maximum may become critical as ΔT increases. Reduced growth may result if the temperature remains too long above the level that allows a maximum scope.

Figure 150. Schematic representation of thermal requirements for different life processes that characterize Pacific salmon. (Brett 1960)

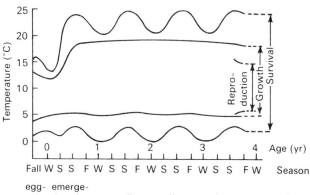

2. Reproduction may be offset from food availability if the ΔT is too high and the optimum temperature for reproduction would occur earlier than the onset of food production, which could be controlled by light intensity.

3. The amount of day-to-day fluctuation that would occur as a result of power-plant problems, flow change, or climate change could result in progressively increasing day-to-day fluctuations with an increasing rise above ambient.

Considering point 3, a hypothetical effect of flow and climate change can be illustrated (Table 51).

Such rapid temperature changes (drops) are possible in heated portions of rivers and would create stress in developing fry; they have been shown to cause warm-water fish, such as smallmouth bass, to desert their nests (Rawson 1945).

The obvious approach with temperature standards, as well as with DO, is to make them less general and more specific for a purpose. The specific purpose would probably be protection of desirable fish species, and the aquatic community in general would be protected by protecting those most sensitive fish (suggested by Mount 1969; Becker 1972; Bush et al. 1974). This seems reasonable if we look again at the position of invertebrates in the species frequency distribution with temperature compared to fish. The lethal temperature mode for freshwater inver-

Table 51. *Effect of flow and climate change on the flow-weighted temperature in a heated portion of a river*

	Flow ($m^3 sec^{-1}$)	°C	Flow-weighted	Mixed °C
Normal conditions during spring reproduction period				
Effluent	2	38	$\dfrac{76 + 84}{6}$	$= 26.7$
River	4	31		
Sudden increase in flow only				
Effluent	2	38	$\dfrac{76 + 252}{14}$	$= 23.4$
River	12	21		
Sudden flow increase and normal river temperature drop				
Effluent	2	38	$\dfrac{76 + 216}{14}$	$= 21.9$
River	12	18		

tebrates is to the right ot that for fish in Figure 151. Based on the assumption that standards to protect fish would also protect the aquatic community in general, such specific guidelines as suggested in Table 52 could be utilized in setting limits for maximum temperature and ΔT in different environments.

With respect to points one and two stated earlier, one should then be able to use these maxima with an ambient temperature cycle, which normally increases gradually from the winter low to the summer high, to judge a safe increment rise above ambient (ΔT) for individual environments. If this is done for the Tennessee River, it can be shown that 2.8 °C is no doubt a safer limit for ΔT than 5.6 °C (Figure 152).

For the Tennessee River, where smallmouth bass approaching 5 kg are a common occurrence, a ΔT rise above ambient of 5.6 °C appears excessive during high-temperature, low-flow years. The periods of reproduction and growth, according to temperature maxima for those activities in Table 52, would occur from 1 to 2 mo earlier and either mortality, substantial weight loss, or a complete evacuation of these

Figure 151. Distribution of freshwater fish (A) and invertebrates (B) according to tolerance to temperature. Dashed line shows approximate mode. (Bush et al. 1974 with permission of the American Chemical Society)

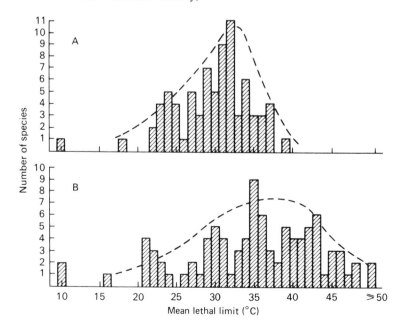

species could be expected in summer even if the maximum is restricted to 34 °C. In fact, Horning and Pearson (1973) have noted that whereas the best growth for smallmouth occurs at 26–29 °C, a temperature even lower is preferred, so 29 or 30 °C may be the maximum tolerable. Interestingly enough, 5.6 °C (10 °F) was accepted by most states in the Tennessee Valley in the late 1960s, but under pressure from the then federal Water Pollution Control Administration, the standard was reduced arbitrarily to 2.8 °C (5 °F). This technique of an optimized ther-

Figure 152. Effect of 2.8 °C and 5.6 °C temperature increments on the normal temperature cycle of the Tennessee River (Welch 1969b). Periods for life-cycle activities that could be affected are shown according to the standards for smallmouth bass and yellow perch. (*Water Quality Criteria* 1968)

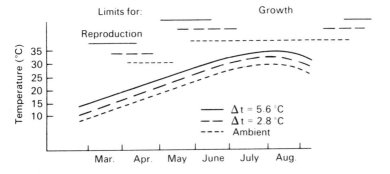

Table 52. *Maximum temperature standard for various life stages of important fish species*

Maximum temperature (°C)	Life stages
34	Growth: catfish, gar, white and yellow bass, buffalo, carpsucker, shad
32	Growth: largemouth bass, drum, bluegill, and crappie
29	Growth: pike, perch, walleye, smallmouth bass, and sauger
26.7	Spawning and egg development of largemouth bass, white and yellow bass, and spotted bass
20	Growth or migration routes of salmonids and for egg development of perch and smallmouth bass
12.8	Spawning and egg development of salmon and trout
9	Spawning and egg development of lake trout, walleye, northern pike, sauger, and Atlantic salmon

Source: *Water Quality Criteria* (1968).

mal regime based on a consideration of the various life-stage requirements has also been illustrated in *Water Quality Criteria* (1973).

Toxicants and toxicity
The variety and quantity of toxicants produced by society increases yearly. Although the toxicity of these compounds can be analyzed on a short-term basis with one or two species, adequate analysis
of the more important long-term, sublethal effects is usually too time
consuming and costly to keep pace with production. In addition, the
variation of the species to be protected in a given environment and the
differing degree of effect under different water-quality characteristics
can be profound. As a result, acceptable concentration limits for even
the more common toxicants are usually not generalized in water-quality
standards. What is usually offered is a statement to the effect that "no
contaminant can be added to water courses that is detrimental to fish
and aquatic life or other water uses."

Determination of toxicity
Controlling waste loading to waters to prevent detrimental effects without specific standards usually necessitates some type of investigation into the toxicity existing in the water or the potential toxicity from wastes either added already or planned for addition. The
investigation usually entails a toxicity bioassay of the waste, the receiving water, and an experimental animal, most often a fish. The
results from such a bioassay provide information about the acute, or
short-term, toxicity, from which a lethal concentration can be determined.

Procedures for acute toxicity bioassays are given in American Public
Health Association (1971). Typically, the exposure period is 2 to 4 days
to a gradient of concentrations with surviving numbers recorded daily
or for shorter intervals. Results are commonly plotted as either percent
survival with time being held constant or as 50% survival with time
variable; both are plotted against concentration (Figure 153). On the
one hand, the lethal concentration is expressed as a $TL_m{}^{96}$-median
tolerance limit after a 96-hr exposure. The concentration producing
50% mortality is interpolated and can also be termed $LC_{50}{}^{96}$, or lethal
concentration for 50% survival after 96 hours.

The lethal threshold concentration can be estimated from the second
procedure by approximating the level at which the resistance time
becomes long at an additional relatively small change in toxicant concentration.

Bioassays can be either static, with no exchange of test water during the 96-hr period, or continuous flow, entailing several solution changes per day. The continuous-flow procedure is considered more appropriate, because in nature organisms are continuously exposed to renewed solutions. In static laboratory solutions, animals can effectively alter the toxicant concentrations and either detoxify the solutions or cause self-contamination with waste products. The lethal concentrations from continuous-flow experiments are apt to be lower than from static. For example, Brungs (1969) reported a $TL_m{}^{96}$ for Zn and fathead minnows in static tests of 12–13 mg l^{-1} but only 8.4 mg l^{-1} in continuous-flow tests.

Of course, knowing the lethal concentration or the concentration that kills 50% of the fish is not an acceptable goal for a receiving water. Conventionally, an application factor is used to proportionately reduce the lethal level to a level that would produce complete survival of all stages of the animal's life history indefinitely. In other words, multiplication of the lethal concentration by the application factor provides the safe concentration.

The application factor (AF) can vary from 0.5 to 0.01. Although in many instances the AF is arbitrary, there has been a considerable effort, largely within the Environmental Protection Agency, to develop factual AFs by determining the effects of certain toxicants on such life functions as growth, fecundity, hatching success, food conversion, and behavior in 1 to 2 mo continuous-flow experiments. With results on the most sensitive life-history stage, the AF can be estimated by:

$$AF = \frac{\text{long-term } TL_m}{TL_m{}^{96}}$$

Results of such AF determinations are given in Warren (1971) for

Figure 153. Two graphical methods of determining the lethal concentration of a toxicant. $TL_m{}^{96}$ (medium tolerance limit after 96-hr exposure) and LTC (lethal threshold concentration) indicated by broken-line arrows.

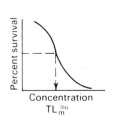
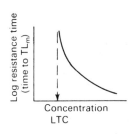

pentachlorophenol, cyanide, dieldrin, 2, 4-D, Malathion, copper, and kraft effluent and range from 0.02 to 0.5 based mostly on reproductive success and growth. Additional values are given in *Water Quality Criteria* (1973) for LAS, chromium, chlorine, sulfides, nickel, lead, and zinc, the latter of which has the lowest AF. As an example of the procedure, Brungs (1969) showed that the growth or hatching success of fathead minnows was not affected by sublethal concentrations of Zn, but that fecundity was (Figure 154). Using the calculated 50% reduction in fecundity and the associated concentration of Zn, the estimated application factor can be calculated by:

$$\frac{0.088 \text{ mg l}^{-1}}{9.2 \text{ mg l}^{-1} \text{ TL}_m{}^{96}} = 0.009$$

Thus, the more conventional AF of 0.1 would be far too large in the case of Zn and fathead minnows.

This procedure is very effective when taking into account differing water-quality characteristics in a receiving water and the complexity of waters differing from source to source. Those variations are presumably accounted for by the lethal-limit determination, which is a simple, relatively quick determination. Although more difficult to obtain, the AF is determined earlier and is then available for use with the lethal-limit determination. There are still two important variables that create problems, however, the extrapolation from one test species to the myr-

Figure 154. Effect of Zn on fecundity in fathead minnows. (Brungs 1969)

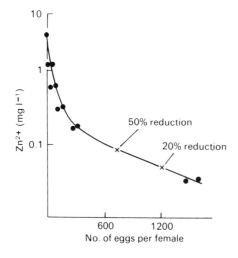

iad of species in the receiving water and the complexity of most wastes, which contain numbers of toxicants that can act together.

Because it is not possible to experiment in the long term with every species in a receiving water, there are three alternatives (reviewed by Woelke 1968): (1) selection of the most sensitive species for AF work, (2) assume proportionality in AFs between an easily studied test species (e.g., fathead minnow) and the most sensitive species in the particular receiving water, or (3) selection of the most economically important species even if it may not be the most sensitive. Woelke (1968) used the reproductive stage of the Pacific oyster and has subsequently shown that it is as sensitive as many other shellfish species to pulp-mill waste as well as oil-spill dispersants and is certainly much more sensitive than fish. In that case, the oyster is economically important as well as sensitive.

Combinations of toxicants present difficulties because toxicants can display either additive, antagonistic, or synergistic effects (Figure 155). If strictly additive, there is no difference in the TL_m whether the mixture is mostly compound A or B. If antagonistic in their effect, of course, the more equal the composition of A and B, the higher the TL_m (or the lower the toxic effect). Antagonism can be shown, for example, between Zn^{2+} and S^-, where toxicity of the mixture is least when the stochiometric ratio is one (Hendricks 1978). Synergism results in a lower TL_m (or more toxicity) as the composition of A and B is equalized. Cu and Zn typically cause synergism when mixed together. Of course, there is a no-effect area where the effect is due to either A or B.

To handle the problem of combinations of toxicants, one approach is to assume an additive effect. An additive model has been used to predict the effects of a combination of toxicants on lethality of fish

Figure 155. Relative effect of toxicant combinations on the TL_m.

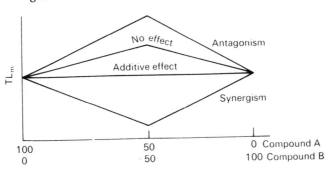

(Brown 1968; Brown et al. 1970). The additive model is:

$$\frac{C_1}{TL_{m(1)}^{96}} + \frac{C_2}{TL_{m(2)}^{96}} + \frac{C_n}{TL_{m(n)}^{96}} = 1.0$$

where $C_1, C_2 \ldots C_n$ are the concentrations of various toxicants in the mixture. Thus, in a mixture where the fractions of all toxicant concentrations, relative to their respective TL_m^{96}, sums to 1.0 or a "total TL_m," 50% mortality should result. In nature, Brown et al. (1970) found that the additive model underestimated the acute toxicity in two polluted rivers and an estuary, that is, the respective fractions produced mortality when the sum was less than 1.0 – on the order of two-thirds. Therefore, one might wish to stipulate that the fractions of the respective toxicants should add up to only less than 1.0, say 0.5–0.8, and a model with which to estimate the long-term safe concentrations in a receiving water could be (*Water Quality Criteria* 1973):

$$\frac{C_1}{TL_{m(1)}^{96} AF_1} + \frac{C_2}{TL_{m(2)}^{96} AF_2} \cdots \frac{C_n}{TL_{m(n)}^{96} AF_n} \leq 1.0 \text{ (e.g. } = 0.8)$$

Effects of specific toxicants
Heavy metals. Metals such as Zn, Cd, Cu, Pb, Cr, Hg, Ag, and Ni are well-known toxic heavy metals that can occur in a variety of wastes and cause either acute or chronic effects on organisms in receiving waters. Although one or more of these metals can originate from a variety of industrial activities, they can also occur in significantly high concentrations in municipal effluents and urban runoff. The recommended maximum concentrations that would likely be safe for aquatic life are shown in Table 53.

An important aspect is the chemical form that the metal takes. One mode of action of metals is to irritate the gills of fishes, causing a secretion of mucus and internal deterioration of the gill lamellae. Fish will die from suffocation as a result; this acute effect will result most readily if the metal is in ionic form. If complexed in inorganic form, either soluble or insoluble, the metal can be relatively nontoxic.

Lloyd (1960) showed that in rainbow trout killed by exposure to $ZnSO_4$ most of the Zn was contained in gills and mucus, which, however, was produced from the body surface and not the gills. Cytological breakdown of the gill tissue was observed within a few hours at high concentrations. Tissue hypoxia resulting from cytological breakdown of the gills has been further demonstrated as the cause for acute toxicity from zinc (Burton et al. 1972).

Over the long term, however, metals are accumulated within body tissues and, in general, uptake and accumulation of a metal at low concentration can occur more readily if it is complexed in organic form and relatively non-ionic. An example of the relative effectiveness of an ionic and non-ionic form was shown with Hg. Ethyl mercury phosphate (EMP) showed a great difference in acute toxicity to fish if added with or without Cl^- (Amend et al. 1969):

		Percent mortality
EMP + H_2O	$CH_3CH_2Hg^+$	0–5
EMP + NaCl	CH_3CH_2HgCl	25–60
EMP + $CaCl_2$	CH_3CH_2HgCl	15–65

With Cl^- added, a unionized organic mercury complex was produced, which would be taken up more readily and therefore produce a greater mortality.

Compounds of intermediate water solubility seem to be most toxic, that is, produce an effect at a lower concentration. If a compound is

Table 53. *Recommended maximum concentrations for representative toxicants; values in ( ) are estimates, largely conservative*

Toxicant	Maximum, mg l^{-1}	AF
Metals		
Al	0.1	—
Cd	0.004 (soft water)	—
Cr	0.05	—
Cu	(0.01)	0.1
Pb	0.03	0.03
Hg	0.00005	—
Ni	(0.1)	0.02
Zn	0.005	0.005
Other inorganics		
NH_3	0.02	0.05
Cl_2	0.003	—
H_2S	0.002	—
HCN	0.005	0.05
Organics		
Phenol	0.1	0.05
LAS (detergent)	0.2	0.05
Organo Cl pesticides	0.002–0.01	—

Source: Water Quality Criteria (1973).

very water soluble (polar), it is readily available, and if in high enough concentration, it can produce an irritation effect on gills. However, the ionized form does not move across lipid membrane surfaces as readily as nonpolar, low-water-soluble compounds. However, if of low-water solubility, it will not be as readily available for uptake. Because of the ease of uptake of organic complexes, high body concentrations have been shown to result from very low concentrations in water.

One of the most toxic compounds is the organic Hg complex, methyl Hg (CH_3Hg). This compound can be produced in nature through microorganism-mediated reactions in the sediment regardless of the form in which Hg is added. Once an environment is contaminated with Hg, the sediments through the continued microbial production of CH_3Hg can contaminate fish for many years. It would not be unusual for a receiving water to require from 20 to 30 years to recover from the detrimental effects of continued recycling of CH_3Hg from sediments to organisms even after the Hg input has ceased (Håkanson 1975; Shin and Krenkel 1976). Although organisms will lose Hg from tissues once the absorption rate has decreased, with a half-life body decay of about 2 yr, CH_3Hg nevertheless can be concentrated in tissue by a factor of thousands of times from water concentrations as low or lower than 1 $\mu g \ l^{-1}$ (Peakall and Lovett 1972). The sediment effect of CH_3Hg production could last from 10 to 100 yr, but normally the sedimentation process may cover the Hg reserve sufficiently to result in recovery to acceptable levels in organisms in from 15 to 30 yr (Håkanson 1975).

Other inorganic toxicants. Examples of common nonmetal toxic inorganic compounds include ammonia (NH_3), hydrogen cyanide (HCN), hydrogen sulfide (H_2S), and chlorine (Cl_2). All of these toxicants are most toxic in their undissociated, molecular state.

The undissociated fraction (NH_3) of ammonia is a function of pH, as shown in Figure 156, because ammonia occurs mostly as ammonium hydroxide and dissociates accordingly:

$$NH_3 + H_2O \rightleftarrows NH_4^+ + OH^-$$

and the equilibrium depends upon pH. About 1.0 mg l^{-1} of NH_3 is acutely lethal to most species of fish. Concentrations of even total NH_3 seldom reach such levels naturally, unless anaerobic conditions persist and nitrification is impeded. However, values of several milligrams per liter are common in waters polluted by industrial and domestic wastes alike. In fact, secondary treatment plant effluent can approach

NH$_3$ concentrations of 20 mg l^{-1}. The importance of pH is readily seen in Figure 156. At pH 7.0, 200 mg l^{-1} of total NH$_3$ would be necessary for lethality, but at pH 9, less than 3 mg l^{-1} of total NH$_3$ would approach lethality.

The maximum limit for NH$_3$, as the undissociated form, is 0.02 mg l^{-1}, and an AF of 0.05 is considered appropriate as indicated in Table 53.

HCN and H$_2$S are the undissociated forms that are toxic. HCN is strongly controlled by pH as is NH$_3$, but in this case the effect is reversed – the HCN fraction is increased as pH is decreased. H$_2$S also depends on pH, accounting for nearly 100% at pH 6.0 and falling below

Figure 156. Percent of ammonia un-ionized (NH$_3$) as a function of pH. (*Water Quality Criteria* 1973)

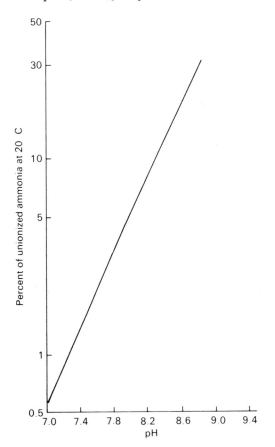

10% as pH approaches 8.0, and is very unstable, as was pointed out in the discussion on the sulfur cycle. It is readily oxidized to SO_4^{2-} in the presence of O_2. Thus, the area affected by H_2S toxicity, where DO is otherwise adequate for aquatic animals, would usually be small. However, a critical condition could exist for developing eggs and larvae near the sediment where H_2S could be generated. Doudoroff et al. (1966) have thoroughly reviewed the toxicity of cyanide. A concentration maximum at any place or time of 0.005 mg l^{-1} (as CN^-) and an AF of 0.05 have been recommended (Table 53).

Chlorine became increasingly important during the early and middle 1970s as investigators discovered that effluents from sewage-treatment plants were toxic to fish because of the residual chlorine they contained. The toxicity of free and/or combined chlorine, often referred to as total residual chlorine ($TRCl_2$), is also controlled by pH. The most toxic form, hypochlorous acid (HOCl), increases as pH decreases. The problem was reviewed by Brungs (1973), who cited work in Michigan with caged fish held downstream from treatment-plant effluents. When the effluent was not chlorinated, no mortality resulted, but when chlorinated, the $TL_m{}^{96}$ was from 0.014–0.029 mg l^{-1} $TRCl_2$. Massive fish kills resulted from effluent chlorine residuals as high as 2.2 mg l^{-1} in the lower James River (Bellanca and Bailey 1977).

Tsai (1973) also investigated 149 sewage-treatment plants. He found fishless waters where the total chlorine content was 0.37 mg l^{-1} and greater, and the observed levels went as high as 1.5 mg l^{-1}. Recent hematological work shows that 0.003 mg l^{-1} is the maximum $TRCl_2$ concentration that would be safe for coho salmon (Buckley et al. 1976; Buckley 1977).

Organic toxicants. The myriad of synthetic organic chemicals produced annually probably pose the major threat, as well as challenge, to man's ability to preserve environmental quality from effects of toxicity. To a large extent this is because of the greater mobility of the more neutrally charged organic molecules to pass across cellular membranes. This fact, together with a great resistance to degradation by many compounds, creates the problem of long-lasting residues and food-chain magnification that has rendered populations of fish unfit for human consumption.

The most famous toxicant in this regard is DDT and its metabolites DDE and DDD. Because it is fat soluble but relatively insoluble in water (~ 1 μg l^{-1}), it can be absorbed by organisms from water or in food resulting in body residues that are 10^6 times the water concentration (Reinart 1970). This absorption can be directly from water and need not involve food transfer, but the two pathways are not additive (Chadwick and Brocksen 1969). Although other organochlorine insecticides are famous for their toxicity and are even more toxic to fish than DDT (e.g., endrin and dieldrin), the overall importance of DDT has been greatest, because it was most heavily used.

As indicated in Table 53, the recommended maximum concentrations of organochlorine insecticides are from 0.002 to 0.01 mg l^{-1}, depending upon the compound. Although such concentrations are one to two orders of magnitude below the lethal concentration, they must be considered liberal, based on the concentrating capability of these compounds in organisms. Fish in Lake Michigan apparently concentrate DDT and dieldrin from the part per trillion level to the part per million level (Reinart 1970).

The most detrimental effect of these compounds, at sublethal (to adults) concentrations, is on the reproduction of fish. This was first demonstrated by Johnson (1967) with the medaka (*Oryzias latipes*). He showed that whereas adults exposed to endrin concentrations of 0.3 μg l^{-1} or less survived in apparent good health, the hatching fry from those fish showed severe behavioral changes and poor survival, which were directly related to the concentration and time of adult exposure. This was true even though the eggs were incubated in endrin-free water.

Although the use of the organochlorine compounds has been greatly reduced or banned altogether, except for special cases, another group of chlorine compounds has become prevalent in the environment. These are the PCBs, or polychlorinated biphenols. These compounds are produced by industry for use as dielectric fluids in capacitors and heat exchangers. The effects, reviewed in some detail in *Water Quality Criteria* (1973), appear very similar to those caused by insecticides. Whereas a maximum water concentration of 0.002 μg l^{-1} is recommended, 0.5 mg kg^{-1} is recommended as a limit for fish tissue.

In addition to the shorter-lived and generally less harmful pesticides that have replaced the organochlorine and organophosphorus types, there are, of course, many other organic toxicants in waste waters. Phenols and detergents (LAS) are quite prevalent in effluents and are rather toxic (Table 53). Fortunately, they do not persist for long and apparently have not caused residue problems.

The herbicides, also acutely toxic, have not caused as severe residue problems as the organochlorine pesticides. The herbicide 2, 4-D is used extensively as an aquatic weedicide as well as in agriculture. It may persist in certain trophic levels and sediment for one to three months but apparently has not caused any detrimental toxic effects to aquatic organisms (Wojtalik et al. 1971). The herbicide 2,4,5-T, however, contains a contaminant dioxin that is highly toxic and presents a residue problem (Galston 1979).

In addition to the problem of residual chlorine, the chlorination of effluents for disinfection results in the production of chlorinated organic compounds that may be detrimental to fish. Chloroform is the most abundant compound and has been considered to have most significance in domestic water supplies with respect to human health (Simler 1978). Toxic chlorinated compounds derived from resin acids in kraft pulp mills are the cause of most of that waste-water toxicity (Leach and Thakore 1975). Five compounds identified had toxicities ranging from 0.32 to 1.5 mg l^{-1} (as the LC$_{50}$^{96}).

Effects of other water-quality characteristics

In addition to synergism and antagonism among more than one toxicant either increasing or decreasing the effects of a given toxicant, water-quality characteristics such as DO, pH, temperature, and water hardness (calcium and magnesium) can also cause such an effect. The role of DO is one of increasing the toxicity. That is, with such toxicants as phenol, Pb, Zn, Cu, and NH$_3$, the effect varies inversely with the DO level. Pickering (1968), for example, showed a lowering in the TL$_m$^{96} for bluegill sunfish exposed to zinc; a TL$_m$^{96} of 10–12 mg l^{-1} at a DO level of 5.6 mg l^{-1}, while at 1.8 mg l^{-1}, the TL$_m$^{96} was near 7 mg l^{-1}.

Usually, the effect of decreasing pH and increasing hardness is to, respectively, increase and decrease toxicity in response to an increase in the concentration of a metal. For example, Cairns et al. (1971c) showed mortality of only 0–10% at pH from 7.3–8.8, whereas complete mortality resulted at pH 5.7–7.0 over Zn concentrations from 10–32 mg l^{-1}. However, Mount (1966) found that toxicity of zinc to fathead minnows increased with increasing pH, which was thought to be an effect of continuous flow bioassay systems keeping precipitated zinc in suspension. The TL$_m$^{96} values increased by a factor of nearly 3 when pH was decreased from 8.6 to 5. In general, however, decreasing pH results in increased ionic fractions of heavy metals, which are the most active toxic fraction. Hardness, on the other hand, increased the TL_m^{96} by more than twofold when increased from 50 to 200 mg l^{-1}. The mechanism that causes hardness to decrease toxicity of a metal is not entirely clear, but may well be physiological rather than chemical solubility. The effect of pH on the toxicity of NH_3, HCN, and H_2S has already been discussed, but in general the relationship is direct with ammonia and inverse with cyanide and sulfide.

The direct effect of pH on fish survival should also be mentioned in conjunction with the problem of acid rain, which is now prevalent in parts of Scandinavia, the eastern United States, and Canada. Rain from air polluted with H^+ and SO_4^{2-} falls on the bedrock areas of Norway and western Sweden and has resulted in many lakes and rivers below pH 5.0 with SO_4^{2-} rather than HCO_3^- as the major ion. The low pH has caused the elimination of salmon from several rivers in southern Norway and of the sport fishery from hundreds of lakes (Wright et al. 1976). The effect of acidity is first felt on fish reproduction, with limits for normal salmon (Atlantic) reproduction being 4.5–5.0 (Jensen and Snekvik 1972). As a result, the fishery gradually declines with an increase in the average age of fish until the fishery is extinguished. Results of catch from an affected river versus the average catch from unaffected rivers in Norway are shown in Figure 157.

Figure 157. Salmon-catch records for 79 Norwegian rivers unaffected by acid rain and for a southern Norway acidic river during 1880–1970. (Wright et al. 1976)

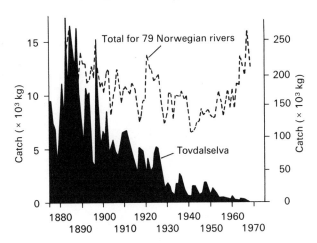

Study questions

1. A brook trout population was observed to gradually disappear as a lake became highly eutrophic. How could the disappearance be explained if the lake was stratified in summer, the temperature in the epilimnion reached an average of 25 °C for a period of several weeks, and the oxygen content in the epilimnion was always super-saturated in the day but did not go below the minimum "standards" at night?

2. If the mean daily temperature was raised from a maximum of 20 °C to 25 °C in a section of stream that contained rainbow trout and smallmouth bass, how would you expect the following to change and why (that is, increase, decrease, or show no change)?
   a. production of trout
   b. production of bass
   c. number of benthic invertebrate species
   d. production of benthic invertebrates

3. Could heat be added to a stream to increase trout production? What maximum temperature increase would you impose to stay within what maximums and when, if the stream naturally ranged between 0 and 12 °C (winter to summer)?

4. The authors of the "Blue Book" (*Water Quality Criteria*, 1973) recommend that for maximum protection of fish and other aquatic animals the minimum dissolved oxygen content in any season should not be less than the natural, which is assumed to be saturation. Why did they recommend that?

APPENDIX A

BENTHIC MACROINVERTEBRATES

The following is a brief account of the taxonomy and biology of the macroinvertebrate fauna that inhabit fresh and to some extent brackish water. For more details see Usinger (1956).

A. Body with 3 pairs of legs (Insecta); abdomen with 2 or 3 long taillike cerci; thorax and abdomen usually with plate-, feather-, tassel-, or fingerlike tracheal gills
  1. Mayflies (Ephemeroptera)
     a. Taxonomy: tarsi with 1 claw; gills plate-, feather- or tassellike and present on one or more of first abdominal segments
     b. Food habits: herbivorous grazers and detrital feeders
     c. Habitat: depositing substrata - Ephemeridae (burrowers); eroding substrata - Heptageniidae (clingers); eroding and depositing substrata - Baetidae (sprawlers on bottom, agile, free ranging and trash and moss inhabiting)
     d. Life history: 3-stage cycle (egg → naiad → adult); usually < 1-1 yr
  2. Stone flies (Plecoptera)
     a. Taxonomy: tarsi with 2 claws; gills usually present and fingerlike and may occur on abdomen, thorax, or labium
     b. Food habits: herbivorous mostly, some carnivorous (Perlidae, Perlodidae)
     c. Habitat: mostly swift currents ($\geq$ 1 ft sec^{-1}) - eroding substrata
     d. Life history: 3-stage cycle; 1-3 yr

B. Body wormlike, elongate and cylindrical to short and obese, or flattened and oval with appendages and sclerotized head capsule
  1. Caddis flies (Trichoptera)
     a. Taxonomy: thorax with 3 pairs of legs, and segment with a pair of hook-bearing appendages; antennae inconspicuous, 1-segmented
     b. Food habits: herbivorous (including detrital) and carnivorous (Rhyacophilidae)
     c. Habitat: usually eroding substrata (free living, net spinners, and case builders); depositing substrata (some case builders)
     d. Life history: 4-stage cycle (egg → larvae → pupae → adult); 1–2 yr
  2. True flies (Diptera)
     a. Taxonomy: thorax without legs; head capsule distinct, at least anteriorly; abdomen with gills or a breathing tube at posterior end
     b. Food habits: detrital herbivorous feeders
     c. Habitat: eroding substrata (cling by permanent attachment); vegetation (cling with hooks); depositing substrata (tube builders)
     d. Life history: 4-stage cycle; < 1–1 yr
C. Body not wormlike, wing rudiments as external flaplike appendages, mouthparts consisting of long and scooplike labium, which covers other mouthparts when folded
  1. Dragonflies and damselflies (Odonata)
     a. Food habits: all predaceous
     b. Habitat: depositing substrata and vegetation (sprawlers, climbers, and burrowers)
     c. Life history: 3-stage cycle; 1–5 yr
D. Animals living within hard carbonate shell: *Mollusca (snails and clams)*
  1. Snails (Gastropoda): shells entire, usually spiral, but may be "coolie-hat" shaped
     a. Food habits: mostly detrital feeders and grazers
     b. Habitat: mainly in depositing substrata, but also in eroding substrata
     c. Life history: 3-stage cycle, < 1–1 yr
  2. Clams, bivalves (Pelecypoda): shells consisting of two hinged halves
     a. Food habits: detrital and herbivorous filter feeders

    b. Habitat: mainly depositing substrata

    c. Life history: 3-stage cycle; < 1-1 yr

E. Body wormlike and divided into many small segments much wider than long; bristles frequently present

  1. Worms and leeches (Annelida)

    a. Food habits: mostly detrital feeders, some predators and parasites

    b. Habitat: usually depositing substrata

    c. Life history: 2- or 3-stage cycle; < 1-1 yr

F. Body not wormlike, with at least 5 pairs of legs: Crustacea

  1. Body flattened laterally: Amphipods (scuds)

    a. Food habits: detrital

    b. Habitat: eroding and depositing substrata with vegetation

    c. Life history: 3-stage cycle; < 1 yr

  2. Body flattened horizontally: Isapoda (aquatic sow bugs), eyes not on stalks; Decapoda (crayfish), eyes on stalks; biology similar to Amphipoda

APPENDIX B

ANSWERS TO STUDY QUESTIONS

Chapter 3. Hydrographics characteristics

1. The flow could be either laminar or turbulent. The turbulent flow is the dominating one in nature because laminar flow can only exist when the water velocity is very low and the bottom is very smooth. In nature all kinds of disturbances exist.

2. The regimes of flow are shooting and streaming. The flow is shooting when the velocity is higher than the velocity of propagation of a gravity wave and streaming when it is lower.

3. The density differences in water caused by temperature, salinity, and suspended material. In rivers usually gradients are caused by the fact that the river erodes and transports material which will appear asymmetrical in a vertical section of the river with higher concentrations closer to the bottom layers. In lakes usually the warming of the water from *the air* causes the surface water to be warm and less dense. To some extent salinity and suspended matter also can cause density gradients close to inflow of rivers. In estuaries the inflowing river water is usually much less dense than the sea water with high salinity.

4. An increase in density gradients within a lake diminishes the exchange of water in the vertical direction.

5. The Richardson number is a way to describe stability conditions with respect to density gradients and velocity of a water mass. If the water mass has a homogeneous density, the Richardson number is 0, and the stability is said to be neutral. For negative values of the number, there are unstable conditions; for positive Richardson numbers, the stratification is stable.

6. A water current may be limited in space by density differences between the inflowing water and the receiving water. Usually those differences are caused by differences in salinity. When the sediment load of an inflowing water course is extremely high, the water with the sediment can continue as a rapid undercurrent along the bottom, using its kinetic energy to erode material from the bottom and thus increase the load and density difference.

7. Starting at about the beginning of the hydrologic year (October) the water of a mid-latitude lake is homogeneous from the surface to the bottom. With increasing losses of energy during continuous mixing the lake is cooled down to about 4 °C. With still more loss of energy, ice is formed at the surface. For the rest of the winter the lake has a temperature between 0–4 °C. The increasing insolation in the spring causes breaking up of the ice and a new circulation of the whole water column. In the summer, surface heating dominates and creates a stable stratification for sufficiently deep lakes. Those lakes are called dimictic.

8. Other types of temperature regimes are called warm and cold monomictic, polymictic, and oligomictic. Monomictic lakes have only one period of complete vertical circulation per year. Polymictic lakes circulate often, and oligomictic lakes circulate sporadically.

9. Surface seiches and internal seiches.

10. The necessary conditions are that the lake is stratified and that the wind causes a displacement of water on the windward side of the lake, which presses down the thermocline. When the wind ceases, the seiche appears.

11. The uninodal surface seiche:

$$T_1 = \frac{2\,l}{\sqrt{g \cdot \bar{z}}} = \frac{2 \cdot 2000}{\sqrt{9.81 \cdot 15}} = 329.8 \approx 5 \text{ min}$$

the internal seiche is:

$$T = \frac{2\,l}{\sqrt{g(\rho_h - \rho_e)\Big/\left(\dfrac{\rho_h}{h_h} + \dfrac{\rho_e}{h_e}\right)}} = \frac{2 \cdot 3000}{\sqrt{\dfrac{9.81(0.9998088 - 0.9988022)}{0.9998088/10 + 0.9988022/5}}}$$

$$= \frac{6000}{\sqrt{\dfrac{9.81 \cdot 0.0010066}{0.29974132}}} = \frac{6000}{\sqrt{0.03294}} = \frac{6000}{0.1815} = 33\ 056 \approx 9.2 \text{ hr.}$$

12. The inertia circle of a lake is defined as $r = u/f$ where $u =$ velocity of the water and $f =$ Coriolis parameter. If the width of a lake is $>5\ r$, you can expect some influence of the rotation of the earth upon the mean current pattern of the lake. If the width is $>20\ r$ then the Coriolis force dominates the currents in the lake.

13. A thermal bar is a very strong horizontal temperature gradient in a lake. It is especially dominant in large lakes in the spring time, but may occur in the fall. It is caused by the quick heating of the coastal waters in comparison to the open waters of the lake.

14. There are very often larger differences in the densities in the vertical than in the horizontal direction, which is why the vertical diffusion is more limited. Vertical diffusion may be very restricted especially in the presence of a thermocline.

15. The dispersion, for instance, of a pollutant in a lake may occur by three different processes. First, complete mixing may occur, with both vertical and horizontal components. Second, inflowing water may disperse mainly in the horizontal direction in the form of overflow, interflow, or underflow. Third, there may be a combination of vertical and horizontal mixing but restricted only to a certain depth interval; e.g. complete mixing within the epilimnion of a lake.

Chapter 4. Population growth

1. Reduce concentration in inflow because the visual quality is related to biomass, which in culture is a function of the concentration. Increasing flushing rate with same inflow concentration would not lower biomass until near washout rate, which would be impractical to reach in a lake with a 1 yr^{-1} flushing rate.

2. b

3. c

4. c

Chapter 6. Nutrient cycles

1. The shallow lake because the volume-to-area ratio is smaller, which would give higher concentration of released P from sediment. There is more chance for wind disturbance and vertical transport of P during the stratified period. The period of an-

aerobiosis is apt to be longer because the hypolimnetic O_2 quantity is less. A deep lake sustains more complete loss back to sediment of released P and occurs only once or twice.

2. a. Phosphorus
   b. Yes or no
   c. Depends on concentration; if relatively low then yes, if high then no. μ may be maximum if P concentration is high, and the ratio would still indicate P to be the most limiting.
3. c
4. c
5. c
6. b
7. c
8. h

Chapter 7. Phytoplankton and controlling factors and Chapter 8. Zooplankton

1. Yes, grazing would allow such a relatively low biomass to produce that much because recycled nutrients would allow the higher growth rate.
2. No; if P is the most limiting nutrient then light is limiting, as indicated by the ratio of photic zone depth to thermocline depth. Because of the low algal content in A relative to B, the greater light attenuation in A must be related to nonalgal suspended matter.
3. CO_2 is available from the atmosphere, even though at a slower rate than required for μ_{max}, in sufficient amount to ultimately produce a biomass in proportion to available P. In the course of μ limitation, from reduced free CO_2 and increased pH, blue-green algae are considered to be favored by a higher μ at low CO_2 concentration than green algae.
4. *Follow* controls on external P inputs, otherwise the lake will quickly reestablish an equilibrium P content, as the P-depleted water is replaced with P-rich water. If inlake control follows external controls, then reduction in inlake concentrations can be expected to have more than a temporary effect.
5. Change in P concentration could agree with the model, but the extent of change may not have been enough to reach, or drop below, critical limit (in concentration) for a given trophic state.

6. Phytoplankton would be dominated by blue-greens, which are too large and chemically inhibit the zooplankton grazers. Phytoplankton must be shifted to greens and diatoms. Planktivorous fish hold down zooplankton, particularly the large species. Such predation effects on grazers must be minimized by maximizing piscivorous fish populations.

7. Because the colonial and filamentous forms, typical of blue-greens, are not readily eaten by zooplankton; thus the large zooplankton that can effectively utilize phytoplankton, increase if the blue-greens do not replace the edible phytoplankton. Thus, more of the phytoplankton are converted to large zooplankton, which are more readily eaten by fish.

8. Because flushing rate (ρ) was added. The total mass entering per year is not as important as the concentration, and correcting the loading for flushing rate essentially gives the average inflow concentration. Another way of saying it is that ρ accounts for the P that leaves the system and cannot be used by algae or cannot be measured as concentration. A shallow lake with a moderately high loading could be above L_c in the old formulation, but well below if it also had a high flushing rate.

10. Mr. Anticarbon, because while C could limit productivity in the short term in enriched systems, where the demand imposed from the high levels of P and N is great, in the long term CO_2 will enter from the atmosphere and tend not to limit an eventual accumulation of biomass which is consistent with the amount of P available.

11. $\bar{P} = 50$ μg l^{-1} after equilibrium is reestablished from the continued external loading. This level is more than the eutrophic limit. In spite of inlake removal of P with alum, which may result in a temporary improvement, within 2 years (most water renewed in a year) the lake should have returned to a eutrophic state from the external load alone.

12. d
13. a
14. b
15. b
16. d
17. a
18. a
19. c
20. c

Chapter 9. Periphyton

1. 3. Accrual will be greater at higher velocity if limiting nutrient content is relatively high, because current maintains the nutrient gradient between water and cell.

    1. Although nutrient-transport efficiency from water to the periphyton mat may be low, and therefore growth rate is low, accrual may be relatively high because erosion loss is low at the low velocity.

    2. Velocity has a more detrimental than beneficial effect because erosional loss is high.

2. A has the most production because current stimulates nutrient uptake by heterotrophs first, followed by algae, because with the DOC present at such a high concentration heterotrophs would crowd out algae.

3. e

4. b

5. c

6. b; c is true at any nutrient concentration, but b is more accurate here because with high PO_4 content (50 μg l^{-1}), fixation of P is sufficiently great to exceed erosion so that PO_4 change greatly affects accrual.

Chapter 10. Macrophytes

1. a. When a large fraction of the external P input could be removed annually by harvesting weeds; that is, a high ratio of weed-P : outside-P exists. Of course, lake morphometry must be suitable (shallow enough) for harvesting.

    b.

	Harvest	*2, 4-D*
Advantage	Leaves some plant habitat, no foreign substances, effective immediately, etc.	Low cost, effective immediately, etc.
Disadvantage	More costly, residual disposal, more treatments needed, etc.	Leaves P in lake, some toxicity, may select for resistance, etc.

2. Because the sediments were enriched by the settled plankton over the eutrophic years, which would not readily be depleted upon waste-effluent diversion. As soon as the water cleared,

allowing greater available light, and the sediment nutrient supply remained intact, the macrophytes could flourish.

3. d

Chapter 11. Benthic macroinvertebrates

1. A biological change caused by extreme chemical values persists and can be recognized any time, while the chemical change is short lived; or a biological change can result from any number of chemicals operating additively or synergistically to produce an undesirable effect, while chemical detection requires analyzing numerous substances with a slim likelihood of picking the problem chemical.

2. a. B, increase, because dissolved organic matter causes heterotroph growth.

   b. A, increase, because more nutrients are available in treated waste and less turbidity;

   c. B, increase, because organic matter in untreated waste is greater than treated waste and organic matter is food supply;

   d. B, decrease, because species number decrease from substrate and O_2 intolerance, and biomass increase from organic food;

   e. B, decrease, because stoneflies are highly intolerant of O_2 depletion and organic sediment.

3. A toxic waste; yes, the quantity changed

   1. More species, intolerant of the higher toxicant level, reappeared;

   2. Biomass almost always decreases with a toxicity increase;

   3. There was still enough toxicity to inhibit grazers, so the periphyton built up higher biomass than at the upstream station where grazing was uninhibited.

Chapter 12. Fish

1. Because 25 °C is lethal and the preferred temperature is 14–17 °C, the trout would be forced out of the surface waters into the cooler metalimnion and hypolimnion. The hypolimnion because of the thermal stratification, loses its O_2, as the production of organic matter increases, down to levels that in early stages of eutrophication could be growth-limiting to fish but in later stages is actually lethal.

2. a. Decrease, because 25 °C is lethal to trout.
   b. Increase, because 25 °C is closer to the preferred temperature of small-mouth bass.
   c. Increase, because 25 °C is closer to the preferred temperature for more species than is 20 °C.
   d. Increase, because productivity should increase with temperature assuming food supply does not limit.
3. One could add heat to the stream such that the summer maximum was raised to 17 °C (or a 5 °C increase). As that is the temperature allowing maximum scope for growth, production would rise. Reproduction would not be severely affected at such an increment, but would occur earlier. A higher increment would probably also increase production, as long as the maximum temperature did not go much above 17 °C and the reproductive season did not occur too early, before food was available.
4. Because results with growth of cold- and warm-water fish, under unlimited food ration, decrease progressively as DO is lowered from saturation. The same is true for the size of emerging salmonid fry and swimming ability of salmonids.

REFERENCES

Adams, J. R. 1969. Ecological investigations related to thermal discharges. Pacific Gas and Electric Co. report.

Ahlgren, G. 1977. Growth of *Ocillatoria agardii* Gom. in chemostat culture. I. Investigation of nitrogen and phosphorus requirements. *Oikos 29*:209-24.

Ahlgren, I. 1978. Response of Lake Norrviken to reuced nutrient loading. *Verh. Internat. Verein. Limnol. 20*:846-50.

Ambühl, H. 1959. Die bedeutung der strömung als ökologischer factor. *Schweiz. Z. Hydrol. 21*:133-264.

Amend, D., W. Yasutake, and R. Morgan. 1969. Some factors influencing susceptibility of Rainbow Trout to the acute toxicity of an ethyl mercury phosphate formulation (Timsan). *Trans. Amer. Fish Soc. 98*:419-25.

American Public Health Association. 1971. Standard methods for the examination of water and waste water, 14th ed.

Anderson, J. B., and W. T. Mason, Jr. 1968. A comparison of benthic macroinvertebrates collected by dredge and basket sampler. *J. Water Poll. Cont. Fed. 40*:252-9.

Antia, N. J., C. D. McAllister, T. R. Parsons, K. Stephens, and J. D. H. Strickland. 1963. Further measurements of primary production using a large volume plastic sphere. *Limnol. and Oceanog. 8*:166-83.

Arnold, D. E. 1971. Ingestion, assimilation, survival, and reproduction by *Daphnia pulex* fed seven species of blue-green algae. *Limnol. and Oceanog. 16*:906-20.

Averett, R. C. 1961. Macroinvertebrates of the Clark Fork River, Montana - a pollution survey. Mont. Bd. of Health and Fish and Game Dept. Rept. No. 61-1.

-. 1969. Influence of temperature on energy and material utilization by juvenile coho salmon. Ph.D. thesis, Oregon State University, Corvallis.

Ayles, B. G., J. G. I. Lark, J. Barica, and H. Kling. 1976. Seasonal mortality of rainbow trout (*Salmo gairdnerii*) planted in small eutrophic lakes of Central Canada. *J. Fish Res. Bd.*, Canada, *33*:647-55.

Baalsrud, K. 1967. Influence of nutrient concentration on primary production. In *Pollution and Marine Ecology*, T. A. Olson and F. J. Burgess (eds.), pp. 159-69. John Wiley & Sons, New York.

Bahr, T. G., R. A. Cole, and H. K. Stevens. 1972. Recycling and ecosystem response to water manipulation. Technical Rep. No. 37, Inst. of Water Res., Michigan State University, East Lansing.

Beak, T. W. 1965. A biotic index of polluted streams and the relationship of pollution to fisheries. Advances in *Water Poll. Res., Proc. 2nd Int. Conf. 1*:191-210.

Beck, W. M. 1955. Suggested method for reporting biotic data. *Sewage and Indust. Wastes 27*:1193–7.

Becker, D. 1972. Columbia River thermal effects study: reactor effluent problems. *J. Water Poll. Cont. Fed. 45*:850–69.

Beeton, A. M. 1965. Eutrophication of the St. Lawrence and Great Lakes. *Limnol. and Oceanog. 10*:240–54.

–. and W. T. Edmondson. 1972. The eutrophication problem. *J. Fish Res. Bd.*, Canada, *29*:673–82.

Bellanca, M. A., and D. S. Bailey. 1977. Effects of chlorinated effluents on aquatic ecosystem in the lower James River. *J. Water Poll. Cont. Fed. 49*:639–45.

Bengtsson, L. 1975. Phosphorus release from a highly eutrophic lake sediment. *Verh. Internat. Verein. Limnol. 19*:1107–16.

–. and C. Gelin. 1975. Artificial aeration and suction dredging methods for controlling water quality. Proc. Symposium, the Effects of Storage on Water Quality, Water Res. Cent., Medmenham, England, pp. 314–42.

–., S. Fleischer, G. Lindmark, and W. Ripl. 1975. The Lake Trummen restoration project. I. Water and sediment chemistry. *Verh. Internat. Verein. Limnol. 19*:1080–7.

Bierman, V., V. Verhoff, T. Poulson, and M. Tenney. 1973. Multinutrient dynamic model of algal growth and species competition in eutrophic lakes. In *Modeling the Eutrophication Process*, J. M. Middlebrooks (ed.), pp. 89–109. Utah State University, Water Res. Lab.

Birch, P. B. 1976. The relationship of sedimentation and nutrient cycling to the trophic status of four lakes in the Lake Washington drainage basin. Ph.D. dissertation, University of Washington, Seattle.

Björk, S. 1974. European lake rehabilitation activities. Institute of Limnology report, University of Lund, 23 pp.

–. et al. 1972. Ecosystem studies in connection with the restoration of lakes. *Verh. Internat. Verein. Limnol. 18*:379–87.

Borman, F. H., and G. E. Likens. 1967. Nutrient cycling. *Sci. 155*:474–29.

Born, S. M., T. L. Wirth, J. P. Peterson, J. P. Wall, and D. A. Stephenson. 1973a. Dilutional pumping of Snake Lake, Wisconsin – a potential renewal technique for small eutrophic lakes. Wisc. Dept. Nat. Res., Tech. Bull. No. 66, 32 pp.

–., E. Brick, and J. O. Peterson. 1973b. Restoring the recreational potential of small impoundments: the Marian Millpond experience. Wisc. Dept. Nat. Res., Tech. Bull. No. 71, 20 pp.

Brett, J. R. 1956. Some principles in the thermal requirements of fishes. *Quart. Rev. Biol. 31*:75–87.

–. 1960. Thermal requirements of fish – three decades of study, 1940–70. In *Biological Problems in Water Pollution*, Trans. 2nd Seminar 1959, USPHS, Cincinnati, Ohio, pp. 111–17.

Brezonik, P. L., and G. F. Lee. 1968. Denitrification as a nitrogen sink in Lake Mendota, Wisc. *Environ. Sci. Technol. 2*:120–25.

Brillouin, L. 1962. *Science and Information Theory*, 2nd ed. Academic Press, New York.

Brinkhurst, R. O. 1965. Observations on the recovery of a British River from gross organic pollution. *Hydrobiol. 25*:9–51.

Brock, D. B. 1970. *Biology of Microorganisms*. Prentice-Hall, Englewood Cliffs, N.J.

Brooks, J. L. 1969. Eutrophication and changes in the composition of the zooplankton. In *Eutrophication: Causes, Consequences, and Correctives*. Nat. Acad. Sci., Washington, D.C.

Brown, V. M. 1968. The calculation of the acute toxicity of mixtures of poisons to rainbow trout. *Water Res. 2*:723–33.

–., D. G. Shurben, and D. Shaw. 1970. Studies on water quality and the absence of fish from some polluted English rivers. *Water Res. 4*:363–82.

Brungs, W. 1969. Chronic toxicity of zinc to the Fathead Minnow, *Pimephales promelas* Rafinesque. *Trans. Amer. Fish Soc. 98*:272-9.

–. 1973. Effects of residual chlorine on aquatic life. *J. Water Poll. Cont. Fed. 45*:2180-93.

Bryan, A., K. Marsden, and S. Hanna. 1975. A summary of observations on aquatic weed control methods. B.C. Dept. of Environment, unpublished report, 73 pp.

Buckley, J. A. 1971. Effects of low nutrient dilution water and mixing on the growth of nuisance algae. M.S. thesis, University of Washington, Seattle, 116 pp.

–. 1977. Heinz body hemolytic anemia in coho salmon (*Oncorhynchus kisutch*) exposed to chlorinated waste water. *J. Fish Res. Bd.*, Canada, *34*:215-24.

–., C. M. Whitmore, and R. I. Matsuda. 1976. Changes in blood chemistry and blood-cell morphology in coho salmon (*Oncorhynchus kisutch*) following exposure to sublethal levels of total residual chlorine in municipal waste water. *J. Fish Res. Bd.*, Canada, *33*:776-82.

Burton, D. J., A. H. Jones, and J. Cairns, Jr. 1972. Acute zinc toxicity to rainbow trout (*Salmo gairdnari*): confirmation of the hypothesis that death is related to tissue hypoxia. *J. Fish Res. Bd.*, Canada, *29*:1463-6.

Bush, R. M., E. B. Welch, and R. J. Buchanan. 1972. Plankton associations and related factors in a hypereutrophic lake. *Water, Air and Soil Poll. 1*:257-74.

–., and B. W. Mar. 1974. Potential effects of thermal discharges on aquatic systems. *Environ. Sci. & Technol. 8*:561-8.

Butcher, R. W. 1933. Studies on the ecology of rivers. I. On the distribution of macrophytic vegetation in the rivers of Britain. *J. Ecol. 21*:58-91.

Cairns, J., Jr. 1956. Effects of increased temperatures on aquatic organisms. *Indust. Wastes. 1*:150-2.

–. and K. L. Dickson. 1971a. A simple method for the biological assessment of the effects of waste discharges on aquatic bottom-dwelling organisms. *J. Water Poll. Cont. Fed. 43*:755-72.

–., J. S. Crossman, K. L. Dickson, and E. E. Herricks. 1971b. The recovery of damaged streams. *Assn. Southeastern Biol. Bull. 18*:79-106.

–. 1971c. The effects of pH, solubility, and temperature upon the acute toxicity of zinc to the bluegill sunfish (*Lepomis macrochirus* Raf.). *Trans. Kans. Acad. Sci. 74*:81-92.

Carlson, A. R., and R. E. Siefert. 1974. Effects of reduced oxygen on the embryos and larvae of lake trout (*Salvelinus namaycush*) and largemouth bass (*Micropterus salmoides*). *J. Fish Res. Bd.*, Canada, *31*:1393-6.

–., and L. J. Herman. 1974. Effects of lowered dissolved oxygen concentrations on channel catfish (*Ictalurus punctatus*) embryos and larvae. *Trans. Amer. Fish Soc. 103*:623-6.

Carlson, R. E. 1977. A trophic state index for lakes. *Limnol. and Oceanog. 22*:361-8.

Carpenter, S. R., and M. S. Adams. 1978. The macrophyte tissue nutrient pool of a hard-water eutrophic lake: implications for macrophyte harvesting. *Aquatic Bot. 3*:239-55.

Carr, J. F., and J. K. Hiltunen. 1965. Changes in the bottom fauna of western Lake Erie from 1930 to 1961. *Limnol. and Oceanog. 10*:551-69.

Chadwick, G. G., and R. W. Brocksen. 1969. Accumulation of dieldrin by fish and selected fish-food organisms. *J. Wildl. Mgt. 33*:693-700.

Chen, C. W. 1970. Concepts and utilities of an ecological model. *J. San. Eng. Div. Proc. Amer. Soc. Civil Eng. 96*:1085-97.

Christensen, M. H., and P. Harremoes. 1972. Biological denitrification in water treatment. Rep. 72-2. Dept. San. Eng., Technical University of Denmark.

Chrystal, G. 1904. Some results in the mathematical theory of seiches. *Proc. Roy. Soc. Edinb. 25*:328-37.

Chutter, F. M. 1969. The effects of silt and sand on the invertebrate fauna of streams and rivers. *Hydrobiol. 34*:57-75.

Cocking, A. W. 1959. The effects of high temperature on roach (*Rutilus rutilus*). II. The effects of temperature increasing at a known constant rate. *J. Exp. Biol.* 36:217-36.

Coffey, B. T., and C. D. McNabb. 1974. Eurasian water milfoil in Michigan. *Mich. Bot.* 13:159-65.

Cole, R. A. 1973. Stream community response to nutrient enrichment. *J. Water Poll. Cont. Fed.* 45:1875-88.

Cooke, G. D., R. T. Heath, R. H. Kennedy, and M. R. McComas. 1978. The ffect of sewage diversion and aluminum sulfate application on two eutrophic lakes. Ecol. Res. Ser. EPA, pp. 101.

Coutant, C. C. 1966. Alteration of the community structure of periphyton by heated effluents. Report to AEC for contract AT (45-1)-1830, Battelle Memorial Inst.

-. 1969. Temperature, reproduction, and behavior. *Chesapeake Sci.* 10:261-74.

Cronberg, G. , C. Gelin, and K. Larsson. 1975. The Lake Trummen restoration project. II. Bacteria, phytoplankton, and phytoplankton productivitiy. *Verh. Internat. Verein. Limnol.* 19:1088-96.

Csanady, G. T. 1970. Dispersal of effluents in the Great Lakes. In *Water Research*, Vol. 4, pp. 79-114. Pergamon Press, Elmsford, N.Y.

Cummins, K. W. 1974. Structure and function of stream ecosystems. *Bio. Sci.* 24:631-41.

-., J. J. Klug, R. G. Wetzel, R. C. Petersen, K. F. Superkropp, B. A. Manny, J. C. Wuycheck, and F. O. Howard. 1972. Organic enrichment with leaf leachate in experimental lotic ecosystems. *Bio. Sci.* 22:719-22.

Curtis, E. J., and D. W. Harrington. 1971. The occurrence of sewage fungus in rivers of the United Kingdom. *Water Res.* 5:281-90.

Davis, G. E., J. Foster, and C. E. Warren. 1963. The influence of oxygen concentration on the swimming performance of juvenile Pacific salmon at various temperatures. *Trans. Amer. Fish Soc.* 92:111-24.

Delwiche, C. C. 1970. The nitrogen cycle. *Sci. Am.* 223:136-47.

Dillon, P. J. 1975. The phosphorus budget of Cameron Lake, Ontario: The importance of flushing rate relative to the degree of eutrophy of a lake. *Limnol. and Oceanog.* 29:28-39.

-., and F. H. Rigler. 1974a. The phosphorus-chlorophyll relationship in lakes. *Limnol. and Oceanog.* 19:767-73.

-. 1974b. A test of a simple nutrient budget model predicting the phosphorus concentration in lake water. *J. Fish Res. Bd.*, Canada, 31:1771-8.

Dimond, J. B. 1967. Pesticides and stream insects. Bull. 23, Maine Forest Serv. and the Conservation Foundation, Washington, D.C.

Doudoroff, P., and D. L. Shumway. 1967. Dissolved oxygen criteria for the protection of fish. In *A Symposium on Water Quality Criteria to Protect Aquatic Life*. Amer. Fish Soc., Spec. Publ. No. 4, pp. 13-19.

-., G. L. Leduc, and C. R. Schneider. 1966. Acute toxicity to fish of solutions containing complex metal cyanides in relation to concentrations of molecular hydrocyanic acid. *Trans. Amer. Fish Soc.* 95:6-22.

Droop, M. R. 1973. Some thought on nutrient limitation in algae. *J. Phycol.* 9:264-72.

Dugdale, R. C. 1967. Nutrient limitation in the sea: dynamics, identification, and significance. *Limnol. and Oceanog.* 12:658-95.

Dunst, R. C., S. M. Born, P. D. Uttormark, S. A. Smith, S. A. Nichols, J. O. Peterson, D. R. Knauer, S. L. Serns, D. R. Winter, and T. L. Wirth. 1974. Survey of lake rehabilitation techniques and experiences. Wis. Dept. Nat. Res., Tech. Bull. No. 75.

Edmondson, W. T. 1966. Changes in the oxygen deficit of Lake Washington. *Verh. Internat. Verein. Limnol.* 16:153-8.

-. 1969. Eutrophication in North America. In *Eutrophication: Causes, Consequences, and Correctives*. Nat. Acad. Sci., Washington, D.C.

−. 1970. Phosphorus, nitrogen, and algae in Lake Washington after diversion of sewage. *Sci. 169*:690-1.

−. 1972. Nutrients and phytoplankton in Lake Washington. In *Nutrients and Eutrophication: The Limiting Nutrient Controversy*, G. E. Likens (ed.). Spec. Symposium, *Limnol. and Oceanog. 1*:172-93.

−. 1974. Secondary production. *Mitt. Internat. Verein. Limnol. 20*:229-72.

−. 1978. Trophic equilibrium of Lake Washington. Ecol. Res. Ser., EPA-600/3-77-087, 36 pp.

Egloff, D. A., and W. H. Brakel. 1973. Stream pollution and a simplified diversity index. *J. Water Poll. Cont. Fed. 45*:2269-75.

Ehrlich, G. G., and K. V. Slack. 1969. Uptake and assimilation of nitrogen in microbiological systems. ASTM STP 488, Amer. Soc. Test. Mat., pp. 11-23.

Ellis, M. M. 1937. Detection and measurement of stream pollution. Bull. 22, U.S. Bur. of Fish. *48*:365-537.

Elser, H. J. 1967. Observations on the decline of water milfoil and other aquatic plants: Maryland, 1962-7. Unpublished rept., Dept. Chesapeake Bay Affairs, 14 pp.

Emery, R. M., C. E. Moon, and E. B. Welch. 1973. Delayed recovery in a mesotrophic lake after nutrient diversion. *J. Water Poll. Cont. Fed. 45*:913-25.

Eppley, R. W. 1972. Temperature and phytoplankton growth in the sea. *Fisheries Bull. 70*:1063-85.

Fisher, S. G., and G. E. Likens. 1972. Stream ecosystem: organic energy budget. *Bio. Sci. 22*:33-5.

Fitzgerald, G. P. 1964. The biotic relationships within water blooms. In *Algae and Man*, D. F. Jackson (ed.), pp. 300-06. Plenum Press, New York.

−. 1970. Anaerobic muds for the removal of phosphorus from lake water. *Limnol. and Oceanog. 15*:550-5.

Fjerdingstad, E. 1964. Pollution of streams estimated by benthal phytomicroorganisms. I. A saprobic system based on communities of organisms and ecological factors. *Int. Rev. Geo. Hydrobiol. 49*:63-131.

Fogg, G. E. 1965. *Algal Cultures and Phytoplankton Ecology*. University of Wisconsin Press, Madison, 126 pp.

Forel, F. A. 1895. *Le Leman: monographie limnologique. Tome 2, Mecanique, Chimie, Thermique, Optique, Acoustique*. Lausanne, F. Rouge, 651 pp.

Fox, H. M., B. G. Simmonds, and R. Washbourn. 1935. Metabolic rates of ephemerid nymphs from swiftly flowing and still waters. *J. Exp. Biol. 12*:179-84.

Fox, J. L., P. L. Brezonik, and M. A. Keirn. 1977. Lake drawdown as a method of improving water quality. Ecol. Res. Ser., EPA-600/3-77-005, 93 pp.

Fuhs, G. W., S. D. Demmerle, E. Canelli, and M. Chew. 1972. Characterization of phosphorus limited plankton. In *Nutrients and Eutrophication: The Limiting Nutrient Controversy*, G. E. Likens (ed.). Spec. Symposium, *Limnol. and Oceanog. 1*:113-33.

Gächter, V. R. 1976. Die tiefenwasserableitung, ein weg zur sanierung von seen. *Hydrobiol. 38*:1-28.

Galston, A. W. 1979. Herbicides: a mixed blessing. *Bio. Sci. 29*:85-90.

Gameson, A. L. H., M. J. Barrett, and J. S. Shewbridge. 1973. The aerobic Thames estuary. *Advances in Water Poll. Res.*, pp. 843-50.

Gammon, J. R. 1969. Aquatic life survey of the Wabash River, with special reference to the effects of thermal effluents on populations of macroinvertebrates and fish. Unpublished report, DePauw University, Greencastle, Ind. 65 pp.

Gannon, J. E., and A. M. Beeton. 1971. The decline of the large zooplankton, *Limnocalanus macrurus* Sars (Copepoda: Calanoida), in Lake Erie. *Proc. 14th Conf. Great Lakes Res.*, Internat. Assoc. Gt. Lakes Res., pp. 27-38.

Gaufin, A. R. 1958. The effects of pollution on a midwestern stream. *Ohio J. Sci. 58*:197-208.

-., and C. M. Tarzwell. 1956. Aquatic macroinvertebrate communities as indicators of organic pollution in Lytle Creek. *Sewage and Indust. Wastes 28:*906-24.

Gerloff, G. C. 1975. Nutritional ecology of nuisance aquatic plants. Ecol. Res. Ser., EPA-660/3-75-027, 78 pp.

-., and P. H. Krombholz. 1966. Tissue analysis to determine nutrient availability. *Limnol. and Oceanog. 11:*529-37.

-., and F. Skoog. 1954. Cell contents of nitrogen and phosphorus as a measure of their availability for growth of Microcystis aeruginosa. *Ecol. 35:*348-53.

Gessner, F. 1959. Hydrobotanik. *Die Physiologischen Grundlagen der Pflanzenverbreitung im Wasser. II. Stoffhausholt.* Berlin, VEB Deutscher Verlag der Wissenschaften, 701 pp.

Gilmartin, M. 1964. The primary production of a British Columbia fjord. *J. Fish Res. Bd.,* Canada, 21:505-38.

Gliwicz, Z. M. 1975. Effect of zooplankton grazing on photosynthetic activity and composition of phytoplankton. *Verh. Internat. Verein. Limnol. 19:*1490-7.

-., and A. Hillbricht-Ilkowska. 1973. Efficiency of the utilization of nanoplankton primary production by communities of filter-feeding animals measured in situ. *Verh. Internat. Verein. Limnol. 18:*197-212.

Goldman, C. R. 1960a. Molybdenum as a factor limiting primary productivity in Castle Lake, Calif. *Sci. 132:*1016-17.

-. 1960b. Primary productivity and limiting factors in three lakes of the Alaskan Peninsula. *Ecol. Monogr. 30:*207-30.

-. 1962. Primary productivity and micronutrient limiting factors in some North American and New Zealand lakes. *Verh. Internat. Verein. Limnol. 15:*365-74.

-., and R. G. Wetzel. 1963. A study of the primary productivity of Clear Lake, Lake County, Calif. *Ecol. 44:*283-94.

-., and R. Carter. 1965. An investigation by rapid carbon-14 bioassay of factors affecting the cultural eutrophication of Lake Tahoe, California-Nevada. *J. Water Poll. Cont. Fed. 37:*1044-59.

Goldman, J. C. 1973. Carbon dioxide and pH: effect on species succession of algae. *Sci. 182:*306-7.

-., and E. J. Carpenter. 1974. A kinetic approach to the effect of temperature on algal growth. *Limnol. and Oceanog. 19:*756-66.

-., D. B. Porcella, E. J. Middlebrooks, and D. F. Toerien. 1971. The effect of carbon on algal growth - its relationship to eutrophication. Occasional Paper 6, Utah State University, Water Research Lab. and College of Engineering, April, 56 pp.

Golterman, H. L. 1972. Vertical movement of phosphate in fresh water. In *Handbook of Environmental Phosphorus,* E. J. Griffith, A. M. Beeton, J. Spencer, and D. Mitchell, pp. 509-37. John Wiley & Sons, New York.

Graham, J. M. 1949. Some effects of temperature and oxygen pressure on the metabolism and activity of the speckled trout, Salvelinus fontinalis. *Canadian J. Res., D,* 27:270-88.

Grandberg, K. 1973. The eutrophication and pollution of Lake Päijäne, Central Finland. *Ann. Bot. Finnici. 10:*267-308.

Great Phosphorus Controversy, The. 1970. *Environ. Sci. Technol. 4:*725-6.

Green, G. H., and B. T. Hargrave. 1966. Primary and secondary production in Bas d'Or Lake, Nova Scotia, Canada. *Verh. Internat. Verein. Limnol. 16:*333-40.

Grzenda, A. R., and R. C. Ball. 1968. Periphyton production in a warm-water stream. *Quarterly Bull.,* Mich. Agr. Exp. Sta., Michigan State University 50:296-303.

Haines, T. A. 1973. Effects of nutrient enrichment and a rough-fish population (carp) on a game-fish population (smallmouth bass). *Trans. Amer. Fish Soc. 102:*346-54.

Håkanson, L. 1975. Mercury in Lake Vänern - present status and prognosis. Swedish Environ. Prot. Bd., NLU, Report No. 80, 121 pp.

Hall, D. J., W. Cooper, and E. Werner. 1970. An experimental approach to the production dynamics and structure of freshwater animal communities. *Limnol. and Oceanog.* 15:839-928.

Hall, J. D., and R. L. Lanz. 1969. The effects of logging on the habitat of coho salmon and cutthroat trout in coastal streams. T. G. Northcote (ed.), pp. 355-75, *Sym. on Salmon and Trout in Streams.* H. R. MacMillan Lectures in Fisheries, University of British Columbia, Vancouver.

Halsey, T. G. 1968. Autumnal and overwinter limnology of three small eutrophic lakes with particular reference to experimental circulation and trout mortality. *J. Fish Res. Bd.,* Canada, 25:81-99.

Hamilton, R. D., and J. Preslan. 1970. Observations on heterotrophic activity in the eastern tropical Pacific. *Limnol. and Oceanog.* 15:395-401.

Hartman, R. T., and D. L. Brown. 1967. Changes in internal atmosphere of submerged vascular hydrophytes in relation to photosynthesis. *Ecol.* 48:252-8.

Hasler, A. D. 1947. Eutrophication of lakes by domestic drainage. *Ecol.* 28:383-95.

Hawkes, H. A. 1962. Biological aspects of river pollution. In *River Pollution. Two: Causes and Effects,* L. Klein (ed.), pp. 311-432. Butterworths, London.

-. 1969. Ecological changes of applied significance induced by the discharge of heated waters. In *Engineering Aspects of Thermal Pollution,* F. L. Parker and P. A. Krenkel (eds.), pp. 15-57. Vanderbilt University Press, Nashville, Tenn.

Hays, F. R., and J. E. Phillips. 1958. Lake water and sediment. III. Radiophosphorus equilibrium with mud, plants, and bacteria under oxidized and reduced conditions. *Limnol. and Oceanog.* 3:459-75.

Heinle, D. R. 1969a. Effects of elevated temperature on zooplankton, *Chesapeake Sci.* 10:186-209.

-. 1969b. *Thermal loading and the zooplankton community. Patuxent thermal studies, supplementary reports.* Nat. Res. Inst., Ref. No. 69-8, University of Maryland.

Hellawell, J. M. 1977. Change in natural and managed ecosystems: detection, measurement and assessment. *Proc. R. Soc. Lond. B,* 197: 31-57.

Hellström, B. 1941. *Wind effect on lakes and rivers.* Kungliga Tekniska Högskolan, Avhandling 26, Stockholm.

Hendrey, G. R. 1973. Productivity and growth kinetics of natural phytoplankton communities in four lakes of contrasting trophic state. Ph.D. dissertation, University of Washington, Seattle.

-., and E. B. Welch. 1974. Phytoplankton productivity in Findley Lake. *Hydrobiol.* 45:45-63.

Hendricks, A. C. 1978. Response of *Selenastrum capricornutum* to zinc sulfides. *J. Water Poll. Cont. Fed.* 50:163-8.

Herbert, D., R. Elsworth, and R. C. Telling. 1956. The continuous culture of bacteria, a theoretical and experimental study. *J. Gen. Microbiol.* 14:601-22.

Hester, F. E., and J. B. Dendy. 1962. A multiple-plate sampler for aquatic macroinvertebrates. *Trans. Amer. Fish Soc.* 91:420.

Hillbricht-Ilkowska, A. I. 1972. Interlevel energy transfer efficiency in planktonic food chains. International Biological Programme - Section PH, December 13, 1972, Reading, England.

Hodgson, R. H., and N. E. Otto. 1963. Pondweed growth and response to herbicides under controlled light and temperature. *Weed Sci.* 11:232-7.

Hoehn, R. C., J. R. Stauffer, M. T. Masnik, and C. H. Hocutt. 1974. Relationships between sediment oil concentrations and the macroinvertebrates present in a small stream following an oil spill. *Environmental Letters* 7:345-52.

Horne, A. J., and C. R. Goldman. 1972. Nitrogen fixation in Clear Lake, Calif. I. Seasonal variation and the role of heterocysts. *Limnol. and Oceanog.* 17:678-92.

Horner, R. R. 1978. An investigation of stream periphyton community development in relation to current velocity and nutrient availability. Ph.D. dissertation. University of Washington.

Horning, W. B., II, and R. E. Pearson. 1973. Growth temperature requirements and lower lethal temperatures for juvenile smallmouth bass (*Micropterus dolomieu*). *J. Fish Res. Bd.*, Canada, *30*:1226-30.

Howard, D. L., J. I. Frea, R. M. Pfister, and P. R. Dugan. 1970. Biological nitrogen fixation in Lake Erie. *Sci. 169*:61-2.

Hulbert, S. H. 1971. The nonconcept of species diversity: a critique and alternative parameters. *Ecol. 52*:577-86.

Hutchinson, G. E. 1957. *A Treatise on Limnology*, Vol. 1. John Wiley & Sons, New York.

-. 1967. *A Treatise on Limnology*, Vol. 2. John Wiley & Sons, New York.

-. 1970a. The biosphere. *Sci. Am. 223 (No. 3)*:44-53.

-. 1970b. The chemical ecology of three species of *Myriophyllum* (Angiospermae, Haloragaceae). *Limnol. and Oceanog. 15*:1-5.

Hynes, H. B. N. 1960. *The Biology of Polluted Water*. Liverpool University Press, Liverpool, England.

Imboden, D. M. 1974. Phosphorus model for lake eutrophication. *Limnol. and Oceanog. 19*:297-304.

Isaac, G. W., R. I. Matsuda, and J. R. Welker. 1966. A limnological investigation of water quality conditions in Lake Sammamish. Munic. Metro. Seattle, *Water Quality Series No. 2*, pp. 47.

Javornicky, P., J. Komarek, and R. Ruzieka. 1962. The phytoplankton of the water reservoir of Slapy during 1958-60. Inst. Chem. Technol., Prague, *Technol. Water 6*:349-87.

Jensen, K. W., and E. Snekvik. 1972. Low pH levels wipe out salmon and trout populations in southernmost Norway. *Ambio. 1*:223-5.

Johnson, H. E. 1967. The effects of endrin on the reproduction of a freshwater fish (*Oryzias latipes*). Ph.D. dissertation, University of Washington, Seattle, pp. 136.

Jones, J. R., and R. W. Bachmann. 1976. Prediction of phosphorus and chlorophyll levels in lakes. *J. Water Poll. Cont. Fed. 48*:2176-82.

Kaesler, R. L., E. E. Herricks, and J. S. Crossman. 1978. Use of indices of diversity and hierarchical diversity in stream surveys. In *Biological Data in Pollution Assessment: Quantitative and Statistical Analyses*, K. L. Dickson, J. Cairns, Jr., and R. J. Livingston (eds.), ASTM STP 652, Amer. Soc. Test. Mat., pp. 92-112.

Keating, K. I. 1977. Blue-green algal inhibition of diatom growth: transition from mesotrophic to eutrophic community structure. *Sci. 199*:971-3.

Kennedy, R. H., and G. D. Cooke. 1974. Phosphorus inactivation in a eutrophic lake by aluminum sulfate application: a preliminary report of laboratory and field experiments. Unpublished manuscript, Kent State University (Ohio), Dept. of Biol., 24 pp.

Kerr, P. C., D. F. Paris, and D. L. Brockway. 1970. The interrelation of carbon and phosphorus in regulating heterotrophic and autotrophic populations in aquatic ecosystems. EPA Rep. 16060 FGS 07/70.

Ketelle, M., and P. D. Uttormark. 1971. Problem lakes in the United States. EPA Tech. Rep. 16010 EHR 12171, pp. 282.

Keup, L. E. 1966. Stream biology for assessing sewage treatment plant efficiency. *Water and Sewage Works 113*:411-17.

Kilham, S. S. 1975. Kinetics of silicon-limited growth in the freshwater diatom *Asterionella formosa*. *J. Phycol. 11*:396-9.

King, D. L. 1970. Role of carbon in eutrophication. *J. Water Poll. Cont. Fed. 42*:2035-51.

-. 1972. Carbon limitation in sewage lagoons. In *Nutrients and Eutrophication*, Special Symposium, Vol. I, Amer. Soc. Limnol. and Oceanog., pp. 98–110.

-., and J. T. Novak. 1974. The kinetics of inorganic carbon-limited algal growth. *J. Water Poll. Cont. Fed.* 46:1812–16.

Klein, L. 1962. *River Pollution, Two: Causes and Effects*. Butterworths, London.

Knapp, R. T. 1943. Density currents: their mixing characteristics and their effect on the turbulence structure of the associated flow. University Iowa studies, Eng. Bull. 27, Iowa City.

Knoechel, R., and J. Kalff. 1975. Algal sedimentation: the cause of a diatom–blue-green succession. *Verh. Internat. Verein. Limnol.* 19:745–54.

Kolkwitz, R. and M. Marsson. 1908. Okologie der pflanzlichen Saprobien. *Berichte der Deutschen Botanischen Gesellschaft.* 26a:505–19.

Kormondy, E. J. 1969. *Concepts of Ecology*. Prentice-Hall, Englewood Cliffs, N.J.

Kuentzel, L. E. 1969. Bacteria, carbon dioxide, and algal blooms, *J. Water Poll. Cont. Fed.* 41:1737–47.

Kvarnäs, H., and T. Lindell. 1970. *Hydrologiska studier i Ekoln*. Rapport over hydrologisk verksamhet inom Naturvårdsverkets Limnologiska Undersökning januari–augusti 1969. UNGI Rapport 3, Uppsala.

-. 1973. A study of the interaction between lake water and river water. *Proc. 16th Conf. Great Lakes Res.*, pp. 989–1000.

Lake Erie Report. 1968. A plan for water-pollution control. Dept. of the Interior, Fed. Water Poll. Admin., 107 pp.

Larsen, D. P., and H. T. Mercier. 1976a. Phosphorus retention capacity of lakes. *J. Fish Res. Bd. Canada* 33:1742–50.

-. 1976b. Shagawa Lake recovery characteristics as depicted by predictive modeling. Proceedings of an EPA-sponsored Symposium on Marine, Estuarine and Freshwater Quality, Ecol. Res. Series, EPA-600/3-76-079, pp. 114–37.

.Larsen, D. P., K. W. Malueg, D. W. Schultz, and R. M. Brice. 1975. Response of eutrophic Shagawa Lake, Minnesota, U.S.A., to point-source phosphorus reduction. *Verh. Internat. Verein. Limnol.* 19:884–92.

Laurent, P., J. Garaucher, and P. Vivier. 1970. The condition of lakes and ponds in relation to carrying out of treatment measures. *Advances in Water Poll. Res.* 2:0-23/1-10.

Lawton, G. W. 1961. Limitation of nutrients as a step in ecological control. Algae and metropolitan wastes, R. A. Taft San. Eng. Center, Tech. Rept. W61-3.

Leach, J. M., and A. N. Thakore. 1975. Isolation and identification of constituents toxic to juvenile rainbow trout (*Salmo gairdneri*) in caustic extraction effluents from kraft pulp mill bleach plants. *J. Fish Res. Bd.*, Canada, 32:1249–57.

Lee, G. F. 1970. Factors affecting the transfer of materials between water and sediments. University of Wisconsin Water Res. Center.

Liebmann, H. 1970. Biological and chemical investigation on the effect of sewage on the eutrophication of the Bavarian lakes. *Advances in Water Poll. Res.* 2:III-22/1-7.

Likens, G. E., and F. H. Borman. 1974. Linkages between terrestrial and aquatic ecosystems. *Bio. Sci.* 24:447–56.

Lin, C. 1971. Availability of phosphorus for *Cladophora* growth in Lake Michigan. *Proc. 14th Conf. Great Lakes Res.*, pp. 39–43.

Lindell, T. 1975. Vänern. In *Vänern, Vättern, Mälaren, Hjälmaren – en översikt*. Statens Naturvårdsverk, Publikationer 1976:1, Stockholm.

Lindeman, R. L. 1942. Trophic dynamic aspect of ecology. *Ecol.* 23:399–418.

Lloyd, R. 1960. The toxicity of zinc sulfate to rainbow trout. *Ann. Appl. Biol.* 48:84–94.

Lund, J. W. G. 1950. Studies on *Asterionella formosa* Hass. II. Nutrient depletion and the spring maximum. *J. Ecol.* 38:1–35.

Maciolek, J. A., and M. G. Maciolek. 1968. Microseston dynamics in a simple Sierra Nevada lake-stream system. *Ecol. 49*:60–75.

Macon, T. T. 1974. *Freshwater Ecology*. John Wiley & Sons, New York, 343 pp.

Mann, K. H. 1965. Heated effluents and their effects on the invertebrate fauna of rivers. *Proc. Soc. Water Treat. and Exam. 14*:45–53.

Marcuson, P. E. 1970. Stream sediment investigation progress report. Mont. Fish and Game Dept., Helena.

Margalef, R. 1958. Temporal succession and spatial heterogeneity in phytoplankton. *Perspectives in Marine Ecology*. University of California Press, p. 323.

Margalef, R. 1969. Diversity and stability: a practical proposal and a model of interdependence. Symposium on Diversity and Stability in Ecological Systems, Brookhaven National Lab., Upton, N.Y.

Markowski, S. 1959. The cooling water of power stations: New factor in the environment of marine and freshwater invertebrates. *J. Animal Ecol. 28*:243–58.

Marshall, P. T. 1958. Primary production in the arctic. *J. Cons. Int. Explor. Mer. 23*: 173–7.

Martin, J. B., Jr., B. N. Bradford, and H. G. Kennedy. 1969. Factors affecting the growth of *Najas* in Pickwick Reservoir. TVA, National Fertilizer Development Center Report.

Mason, W. T., Jr., J. B. Anderson, R. D. Kreis, and W. C. Johnson. 1970. Artificial substrate sampling, macroinvertebrates in a polluted reach of the Klamath River, Oregon. *J. Water Poll. Cont. Fed. 42(pt.2)*:R315–27.

McConnell, J. W., and W. F. Sigler. 1959. Chlorophyll and productivity in a mountain river. *Limnol. and Oceanog. 4*:335–51.

McDonnell, J. C. 1975. In situ phosphorus release rates from anaerobic lake sediments. M.S. thesis, University of Washington, Seattle.

McGauhey, P. H., G. A. Rohlich, and E. P. Pearson. 1968. Eutrophication of surface waters – Lake Tahoe: Bioassay of nutrient sources. First Progress Report, Lake Tahoe Area Council, 178 pp.

McIntire, C. D. 1966. Some effects of current velocity on periphyton communities in laboratory streams. *Hydrobiol. 37*:559–70.

–., and H. K. Phinney. 1965. Laboratory studies of periphyton production and community metabolism in lotic environments. *Ecol. Monogr. 35*:237–58.

McMahon, J. W., and F. H. Rigler. 1965. Feeding rate of *Daphnia magna* Straus on different foods labeled with radioactive phosphorus. *Limnol. and Oceanog. 10*:105–13. 13.

Menzel, D. W., and J. H. Ryther. 1964. The composition of particulate organic matter in the western North Atlantic. *Limnol. and Oceanog. 9*:179–86.

Mihursky, J. 1969. Patuxent thermal studies. Summary and recommendations. Nat. Res. Inst. Spec. Rept. No. 1, University of Maryland, College Park, Md.

–., and V. S. Kennedy. 1967. Water temperature criteria to protect aquatic life. In *Symposium on Water Quality Criteria*. Amer. Fish Soc. Spec. Pub. *4*:20–32.

Miller, W. E., T. E. Maloney, and J. C. Greene. 1974. Algal productivity in 49 lake waters as determined by algal assays. *Water Res. 8*:667–79.

Misra, R. D. 1938. Edaphic factors in the distribution of aquatic plants in the English Lakes. *J. Ecol. 26*:411–51.

Mortimer, C. H. 1941. The exchange of dissolved substances between mud and water in lakes (Parts I and II). *J. Ecol. 29*:280–329.

–. 1942. The exchange of dissolved substances between mud and water in lakes (Parts III, IV, summary, and references). *J. Ecol. 30*:147–301.

–. 1952. Water movements in lakes during summer stratification; evidence from the distribution of temperature in Lake Windermere, with an appendix by M. S. Longuet-Higgins. *Phil. Trans.*, Ser. B. *236*:355-404.

Mount, D. I. 1966. The effect of total hardness and pH on acute toxicity of zinc to fish. *Air and Water Poll. Int. J.* 10:49-56.

–. 1969. Developing thermal requirements for freshwater fishes. In *Biological Aspects of Thermal Pollution*, P. A. Krenkel and F. L. Parker (eds.), pp. 140-7. Vanderbilt University Press, Nashville, Tenn.

Mulligan, H. F., and A. Barnowski. 1969. Growth of phytoplankton and vascular aquatic plants at different nutrient levels. *Verh. Internat. Verein. Limnol.* *17*:302-10.

Nebeker, A. V. 1971a. Effect of water temperature on nymphal feeding rate, emergence, and adult longevity of the stone fly *Pteronarcys dorsata. J. Kans. Entomol. Soc.* 44:21-6.

–. 1971b. Effect of high winter water temperatures on adult emergence of aquatic insects. *Water Res.* 5:777-83.

Neil, J. H., and G. E. Owen. 1964. Distribution, environmental requirements, and significance of *Cladophora* in the Great Lakes. *Proc. 7th Conf. Great Lakes Res.* 11:113-21.

Newroth, P. R. 1975. Management of nuisance aquatic plants. Unpublished report, B. C. Dept. of the Environ., 13 pp.

Nichols, D. S., and D. R. Keeney. 1976a. Nitrogen nutrition of *Myriphyllum spicatum*: variation of plant tissue nitrogen concentration with season and site in Lake Wingra. *Freshwater Biol.* 6:137-44.

–. 1976b. Nitrogen nutrition of *Myriophyllum spicatum*: uptake and translocation of ^{15}N by shoots and roots. *Freshwater Biol.* 6:145-54.

Nichols, S. A. 1974. Mechanical and habitat manipulation for aquatic plant management. Wisconsin Dept. Nat. Res., Tech. Bull. No. 77, 34 pp.

–., and G. Cottam. 1972. Harvesting as a control for aquatic plants. *Water Res. Bull.* 8:1205-10.

Nüman, W. 1972. Predictions on the development of the salmonid communities in the European oligotrophic subalpine lakes during the next century. Unpubl. Ms. Staatliche Institut für Seenforschung und Seenbewirtschaftung, Langenargen/ Bodensee, Germany, pp. 12.

Odum, E. P. 1959. *Fundamentals of Ecology.* W. B. Saunders, Philadelphia.

–. 1969. The strategy of ecosystem development. *Sci.* 164:264-70.

Odum, H. T. 1956. Primary production in flowing waters. *Limnol. and Oceanog.* 1:102-17.

Oglesby, R. 1969. Effects of controlled nutrient dilution on the eutrophication of a lake. In *Eutrophication: Causes, Consequences, and Correctives*, pp. 483-93. Nat. Acad. Sci., Washington, D.C.

Ohle, W. 1953. Phosphor als Initialfaktor der Gewässereutrophierung. *Vom Wasser* 20:11-23.

–. 1975. Typical steps in the change of a limnetic ecosystem by treatment with therapeutica. *Verh. Internat. Verein. Limnol.* 19:1250.

Olson, P. R. 1971. Primary production and zooplankton. In *The Fern Lake Studies*, L. R. Donaldson, P. R. Olson, S. Olsen, and Z. Short (eds.), pp. 28-32. University of Washington, Seattle.

Olson, T. A., and M. E. Rueger. 1968. Relationship of oxygen requirements to index-organism classification of immature aquatic insects. *J. Water Poll. Cont. Fed.* 40:188-202.

Ormerod, J. G., B. Grynne, and K. S. Ormerod. 1966. Chemical and physical factors involved in heterotrophic growth response to organic pollution. *Verh. Internat. Verein. Limnol.* 16:906-10.

Patalas, K. 1970. Primary and secondary production in a lake heated by a thermal power plant. *Proc. Inst. Environ. Sci.*, 16th Annual Tech. Meet., pp. 267-71.

Patrick, R. 1966. The effect of varying amounts and ratios of nitrogen and phosphate on algal blooms. *Proc. 21st Indust. Waste Conf.*, pp. 41-51.

-., B. Brum, and J. Cloes. 1969. Temperature and manganese as determining factors in the presence of diatom or blue-green algal floras in streams. *Proc. Nat. Acad. Sci.* 64:472-8.

Peakall, D. B., and R. J. Lovett. 1972. Mercury: its occurrence and effects in the ecosystem. *Bio. Sci.* 22:20-5.

Pearsall, W. H. 1920. The aquatic vegetation of the English Lakes. *J. Ecol.* 8:163-99.

-. 1929. Dynamic factors affecting aquatic vegetation. *Proc. Int. Cong. Plant Sci.* 1:667-72.

Peltier, W. H., and E. B. Welch. 1969. Factors affecting growth of rooted aquatic plants in a river. *Weed Sci.* 17:412-16.

-. 1970. Factors affecting growth of rooted aquatic plants in a reservoir. *Weed Sci.* 18:7-9.

Percival, E., and H. Whitehead. 1929. A quantitative study of the fauna of some types of stream bed. *J. Ecol.* 17:282-314.

Peters, J. C. 1960. Stream sedimentation project progress report. Mont. Fish and Game Dept., Helena.

-. 1967. Effects on a trout stream of sediment from agricultural practices. *J. Wildl. Mgt.* 31:805-12.

Pederson, G. L., E. B. Welch, and A. H. Litt. 1976. Plankton secondary productivity and biomass: their relation to lake trophic state. *Hydrobiol.* 50:129-144.

Peterson, S. A., W. D. Sanville, F. S. Stay, and C. F. Powers. 1976. Laboratory evaluation of nutrient inactivation compounds for lake restoration. *J. Water Poll. Cont. Fed.* 48:817-31.

Phaup, J. D., and J. Gannon. 1967. Ecology of *Sphaerotilus* in an experimental outdoor channel. *Water Res.* 1:523-41.

Phillips, G. L., D. Einson, and E. Moss. 1978. A mechanism to account for macrophyte decline in progressively eutrophicated fresh waters. *Aquatic Bot.* 3:239-55.

Phillips, J. E. 1964. The ecological role of phosphorus in waters with special reference to microorganisms. In *Principles and Applications in Aquatic Microbiology*. H. Heukelekian and N. C. Doudero (eds.), pp. 61-81. John Wiley & Sons, New York.

Phinney, H. K., and C. D. McIntire. 1965. Effect of temperature on metabolism of periphyton communities developed in laboratory streams. *Limnol. and Oceanog.* 10:341-4.

Pickering, Q. H. 1968. Some effects of dissolved oxygen concentrations upon the toxicity of zinc to the bluegill *Lepomis macrochirus* Raf., *Water Res.* 2:187-94.

Pitcairn, C. E., and H. A. Hawkes. 1973. The role of phosphorus in the growth of *Cladophora. Water Res.* 7:159-71.

Pomeroy, L. W. 1974. The ocean's food web, a changing paradigm. *Bio. Sci.* 24:499-503.

Porath, H. A. 1976. The effect of urban runoff on Lake Sammamish periphyton. M.S. thesis, University of Washington, Seattle, 74 pp.

Porter, K. G. 1972. A method for the in situ study of zooplankton grazing effects on algal species composition and standing crop. *Limnol. and Oceanog.* 17:913—17.

Prandtl, L. 1952. *Essentials of Fluid Dynamics*. Blackie & Son, Glasgow.

Rast, W., and G. F. Lee. 1978. Summary analysis of the North American (U.S. portion) OECD eutrophication project: Nutrient Loading - lake response relationships and trophic state indices. Ecol. Res. Ser. EPA-600 13-78-008, 455 pp.

Rawson, D. S. 1945. The experimental introduction of smallmouth bass into lakes of the Prince Albert National Park, Saskatchewan. *Trans. Amer. Fish Soc.* 73:19–31.

–. 1956. Algal indicators of trophic lake types. *Limnol. and Oceanog.* 1:18–25.

–. 1959. Limnology and fisheries of Cree and Wollaston Lakes in northern Saskatchewan. Saskatchewan Dept. of Nat. Res., Fisheries Rept. No. 4.

Raymont. J. E. 1963. *Plankton and Productivity in the Oceans.* Pergamon Press, Elmsford, N.Y.

Reinart, R. E. 1970. Pesticide concentrations in Great Lakes fish. *Pesticides Monitoring J.* 3:233–40.

Resh, V. H., and J. D. Unzicker. 1975. Water quality monitoring and aquatic organisms: the importance of species identification. *J. Water Poll. Cont. Fed.* 47:9–19.

Rickert, D. A., R. R. Petersen, S. W. McKenzie, W. G. Hines, and S. A. Wille. 1977. Algal conditions and the potential for future algal problems in the Willamette River, Oregon. U.S. Geological Circular 715-G, 39 pp.

Rigler, F. H. 1971. Laboratory measurements of processes involved in secondary production – feeding rates of zooplankton. In *Secondary productivity in fresh waters,* W. T. Edmondson, and G. G. Winberg, (eds.), Handbook No. 17, IBP, Blackwell Scientific Publications.

Ringler, N. H., and J. D. Hall. 1975. Effects of logging on water temperature and dissolved oxygen in spawning beds. *Trans. Amer. Fish Soc.,* 104:111–21.

Ripl, W. 1976. Biochemical oxidation of polluted lake sediments with nitrate – a new lake restoration method. *Ambio.* 5:132–5.

Robel, R. J. 1961. Water depth and turbidity in relation to growth of sago pondweed. *J. Wildl. Mgt.* 25:436–8.

Rock, C. A. 1974. The trophic status of Lake Sammamish and its relationship to nutrient income. Ph.D. dissertation, University of Washington, Seattle.

Rodgers, G. K. 1966. The thermal bar in Lake Ontario, spring 1965 and winter 1965–66. Great Lakes Res. Div., University of Michigan, Pub. No. 15, pp. 369–74.

Rodhe, W. 1948. Environmental requirements of freshwater plankton algae. Experimental studies in the ecology of phytoplankton. Symbol. Bot. Upsalien. 10, 149 pp.

–. 1969. Crystallization of eutrophication concepts in northern Europe. In *Eutrophication: Causes, Consequences, Correctives.* Nat. Acad. Sci., Washington, D.C.

Russell-Hunter, W. D. 1970. *Aquatic Productivity.* Macmillan, New York.

Ryding, S., and C. Forsberg. 1976. Six polluted lakes: a preliminary evaluation of the treatment and recovery processes. *Ambio.* 5:151–6.

Ryther, J. H. 1960. Organic production by planktonic algae and its environmental control. In *The Ecology of Algae,* pp. 72–83. Spec. Pub. No. 2, Pymatuning Lab. of Field Biol., University of Pittsburgh.

–., and W. M. Dunstan. 1971. Nitrogen, phosphorus, and eutrophication in the coastal marine environment. *Sci.* 171:1008–13.

–., K. R. Tenore, and J. E. Huguenin. 1972. Controlled eutrophication – increasing food production from the sea by recycling human wastes. *Bio. Sci.* 22:144–52.

Sakamoto, M. 1971. Chemical factors involved in the control of phytoplankton production in the Experimental Lakes Area, northwestern Ontario. *J. Fish Res. Bd.,* Canada, 28:203–13.

Sawyer, C. N. 1947. Fertilization of lakes by agricultural and urban drainage. *J. New England Water Works Assn.* 61:109–27.

Schell, W. R. 1976. Biogeochemistry of radionuclides in aquatic environments. Annual progress report to Energy Res. and Development Admin. RLO-2225-T18-18.

Schelske, C. L., and E. F. Stoermer. 1971. Eutrophication, silica depletion, and predicted changes in algal quality in Lake Michigan. *Sci.* 173:423–4.

Schiff, J. A. 1964. Protists, pigments, and photosynthesis. In *Principles and Applications in Aquatic Microbiology,* H. Heukelekian and N. C. Doudero (eds.), pp. 298–313. John Wiley & Sons, New York.

Schindler, D. W. 1971a. A hypothesis to explain the differences and similarities among lakes in the Experimental Lakes Area, northwestern Ontario. *J. Fish Res. Bd., Canada, 28:*295–301.

–. 1971b. Carbon, nitrogen, and phosphorus and the eutrophication of freshwater lakes. *J. Phycol. 7:*321–9.

–. 1974a. Eutrophication and recovery in experimental lakes: implications for lake management. *Sci. 184:*897–9.

–., and E. J. Fee. 1974b. Experimental lakes area, whole lake experiments in eutrophication. *J. Fish Res. Bd.,* Canada, *31:*937–53.

Schultz, D. W., and K. W. Malueg. 1976. Uptake of radiophosphorus by rooted aquatic plants. Proc. 3rd Nat. Symposium on Radioecology, pp. 417–24.

Schumacher, G. J., and L. A. Whitford. 1965. Respiration and ^{32}P uptake in various species of freshwater algae as affected by current. *J. Phycol. 1:*78–80.

Shannon, C. E., and W. Weaver. 1963. *The Mathematical Theory of Communication.* University of Illinois Press, Urbana.

Shapiro, J. 1970. A statement on phosphorus. *J. Water Poll. Cont. Fed. 42:*772–5.

–. 1973. Blue-green algae: why they become dominant. *Sci. 179:*382–4.

–., V. Lamarra, and M. Lynch. 1975. Biomanipulation: an ecosystem approach to lake restoration. Proc., Symposium on Water Quality Management through Biological Control, U.S. EPA and University of Florida, January, pp. 85–96.

Shin, E., and P. A. Krenkel. 1976. Mercury uptake by fish and biomethylation mechanisms. *J. Water Poll. Cont. Fed. 48:*473–501.

Shumway, D. L., C. E. Warren, and P. Doudoroff. 1964. Influence of oxygen concentration and water movement on the growth of steelhead trout and coho salmon embryos. *Trans. Amer. Fish Soc. 93:*342–56.

Siefert, R. E., and W. A. Spoor. 1973. Effects of reduced oxygen on embryos and larvae of the white sucker, coho salmon, brook trout and walleye. In *The Early Life History of Fish* J. H. S. Baxter (ed.), pp. 487–95. Proc. of Symposium, Marine Biol. Assn., Avon, Scotland.

Silvey, J. K., D. E. Henley, and J. T. Wyatt. 1972. Planktonic blue-green algae: growth and odor-production studies. *J. Amer. Water Works Assn. 64:*35–9.

Simler, J. J. 1978. Mechanistic approach to the formation of trihalogenated methane compounds during the chlorination of water and waste water. Ph.D. dissertation, University of Washington, Seattle.

Simons, T. J. 1973. Development of three-dimensional numerical models of the Great Lakes. Environ. Canada Water Mgt., Scientific Series No. 12.

Skulberg, O. M. 1965. Algal cultures as a means to assess the fertilizing influence of pollution. *Advances in Water Poll. Res. 1:*113–27.

Sládečková, A., and V. Sládeček. 1963. Periphyton as indicator of the reservoir water quality. I. True periphyton. Sci. Pap. Inst. Chem. Technol., Prague, *Technol. Water 7:*507–61.

Smith, G. E., T. F. Hall, Jr., and R. A. Stanley. 1967. Eurasian water-milfoil in the Tennessee Valley. *Weed Sci. 15:*95–8.

Smith, M. W. 1969. Changes in environment and biota of a natural lake after fertilization. *J. Fish Res. Bd.,* Canada, *26:*3101–32.

Smith, S. A., J. O. Peterson, S. A. Nichols, and S. M. Born. 1972. Lake deepening by sediment consolidation – Jyme Lake. Inland Lake Demonstration Proj. Rept., University of Wisconsin and Wis. Dept. of Nat. Res., pp. 36.

Smith, W. E. 1970. Tolerance of *Mysis relicta* to thermal shock and light. *Trans. Amer. Fish Soc. 99:*418-21.

Snodgrass, W. J., and C. R. O'Melia. 1975. A predictive model for phosphorus in lakes. *Environ. Sci. & Technol. 9:*937-45.

Sorokin, C. 1959. Tabular comparative data for the low- and high-temperature strains of *Chlorella. Nature 184:*613-14.

Spence, D. H. N. 1967. Factors controlling the distribution of freshwater macrophytes with particular reference to the lochs of Scotland. *J. Ecol. 55:*147-70.

Stamnes, R. 1972. The trophic status of three lakes related to nutrient loading. M.S. thesis, University of Washington, Seattle.

Stauffer, R. E., and G. F. Lee. 1973. The role of thermocline migration in regulating algal blooms. In *Modeling the Eutrophication Process.* E. J. Middlebrooks, D. H. Falkenborg, and T. E. Maloney (eds.), pp. 73-82. Proc. of Workshop at Utah State University, Logan.

Steele, J. H. 1962. Environmental control of photosynthesis in the sea. *Limnol. and Oceanog. 7:*137-50.

Stewart, K. M. 1976. Oxygen deficits, clarity, and eutrophication in some Madison Lakes. *Int. Revue ges. Hydrobiol. 61:*563-79.

Stewart, N. E., D. L. Shumway, and P. Doudoroff. 1967. Influence of oxygen concentration on the growth of juvenile largemouth bass. *J. Fish Res. Bd.,* Canada, 24:475-94.

Stockner, J. G. 1972. Paleolimnology as a means of assessing eutrophication. *Verh. Internat. Verein. Limnol. 18:*1018-30.

-., and K. R. S. Shortreed. 1978. Enhancement of autotrophic production by nutrient addition in a coastal rain-forest stream on Vancouver Island. *J. Fish Res. Bd.,* Canada, 35:28-34.

Storr, J. F., and R. A. Sweeney. 1971. Development of a theoretical seasonal growth response curve of *Cladophora glomerata* to temperature and photoperiod. *Proc. 14th Conf. Great Lakes Res. 14:*119-27.

Stumm, W. 1963. Proc. of Inter. Conf. on Water Poll. Res., Vol. 2. Pergamon Press, Elmsford, N.Y.

-., and J. O. Leckie. 1971. Phosphate exchange with sediments: its role in the productivity of surface waters. *Proc. 5th Int. Conf. Water Pollut. Res.* III-26/1-16.

Sundborg, A. 1956. The river Klarälven. A study of fluvial processes. Geografiska Annaler Häfte 2-3/1956, pp. 127-316.

Sverdrup, H. U., M. W. Johnson, and R. H. Fleming. 1942. *The Oceans, Their Physics, Chemistry, and General Biology.* Prentice Hall, Englewood Cliffs, N.J.

Sylvester, R. O., and G. C. Anderson. 1964. A lake's response to its environment. *J. San. Eng. Div.,* Amer. Soc. Civil Eng., *SA1:*1-22.

Talling, J. F. 1957. Photosynthetic characteristics of some freshwater plankton diatoms in relation to underwater radiation. *New Phytol. 56:*29-50.

-. 1965. The photosynthetic activity of phytoplankton in East African lakes. *Int. Rev. ges. Hydrobiol. 50:*1-32.

-. 1971. The underwater light climate as a controlling factor in the production ecology of freshwater phytoplankton. *Mitt. Internat. Verein. Limnol. 19:*214-43.

Taylor, E. W. 1966. Forty-second report on the results of the bacteriological examinations of the London waters for the years 1965-6. Metropolitan Water Board, New River Head, London.

Templeton, W. L., C. D. Becker, J. D. Berlin, C. C. Coutant, J. M. Dean, M. P. Fujihara, R. E. Nakatani, P. A. Olson, E. F. Prentice, and D. G. Watson. 1969. Biological effects of thermal discharges. Progress report for 1968. AEC Research and Development Report, BNWL-1050 reprint.

Tenney, M., and W. Echelberger, Jr. 1970. Fly-ash utilization in the treatment of polluted waters. Bur. Mines Info. Circ. No. 8488 Ash Utilization Symp., pp. 236-68.

Thomas, E. A. 1969. The process of eutrophication in Central European lakes. In *Eutrophication: Causes, Consequences and Correctives*. Nat. Acad. Sci., Washington, D.C.

Thut, R. N. 1969. A study of the profundal bottom fauna of Lake Washington. *Ecol. Monogr. 39:79-100*.

Trembly, F. J. 1960. Research project on effects of condenser discharge water on aquatic life. Progress report 1956-9. Inst. of Research, Lehigh University, Bethlehem, Pa.

Tsai, C. 1973. Water quality and fish life below sewage outfalls. *Trans. Amer. Fish Soc. 102:281-92*.

Uhlmann, D. 1971. Influence of dilution, sinking and grazing rate on phytoplankton populations of hyperfertilized ponds and microecosystems. *Mitt. Internat. Verein. Limnol. 19:100-24*.

Usinger, R. L. 1956. *Aquatic Insects of California*. University of Calif. Press, Berkeley.

Uttormark, P. D., and M. L. Hutchins. 1978. Input-output models as decision criteria for lake restoration. Tech. Rep., Wis. Water Res. Center 78-03, 61 pp.

Vallentyne, J. 1972. Discussion in *Nutrients and Eutrophication*, Special Symposium, Vol. 1, *Limnol. and Oceanog.*, p. 107.

Vance, B. D. 1965. Composition and succession of cyanophycean water blooms. *J. Phycol. 1:81-6*.

Vollenweider, R. A. 1968. Scientific fundamentals of the eutrophication of lakes and flowing waters, with particular reference to nitrogen and phosphorus as factors in eutrophication. Paris, Rept. Organization for Economic Cooperation and Development. DAS/CSI/68.27, pp. 182.

-. 1969a. Possibilities and limits of elementary models concerning the budget of substances in lakes. *Arch. Hydrobiol. 66:1-36*.

-., (ed.). 1969b. A manual on methods for measuring primary production in aquatic environments. *Int. Biol. Program Handbook 12*. Blackwell Scientific Publications, Oxford, 213 pp.

-. 1975. Input-output models with reference to the phosphorus-loading concept in limnology. *Schweizerische Zeitschrift für Hydrologie 37:53-84*.

-. 1976. Advances in defining critical loading levels for phosphorus in lake eutrophication. *Mem. Inst. Ital. Idrobiol. 33:53-83*.

-., and P. J. Dillon. 1974. The application of the phosphorus-loading concept to eutrophication research. National Res. Council, Canada. Tech. Rept. 13690, 42 pp.

Warren, C. E., J. H. Wales, G. E. Davis, and P. Doudoroff. 1964. Trout production in an experimental stream enriched with sucrose. *J. Wildl. Mgt. 28:617-60*.

-. 1971. *Biology and Water Pollution Control*. W. B. Saunders, Philadelphia.

Water Quality Criteria (the "Green Book"). 1968. Tech. Advisory Comm. to EPA. U.S. Gov. Printing Office.

Water Quality Criteria (the "Blue Book"). 1973. Tech. Advisory Comm. Nat. Acad. of Sci. and Acad. of Engineers. U.S. Gov. Printing Office.

Welch, E. B. 1969a. Factors controlling phytoplankton blooms and resulting dissolved oxygen in Duwamish River estuary, Washington. U.S. Geol. Survey Water Supply Paper 1873-A, 62 pp.

-. 1969b. Discussion of ecological changes of applied significance induced by the discharge of heated waters. In *Thermal Pollution, Engineering Aspects*, F. L. Parker and P. A. Krenkel (eds.), pp. 58-68. Vanderbilt University Press, Nashville, Tenn.

-. 1977. Nutrient diversion: Resulting lake trophic state and phosphorus dynamics. Ecol. Res. Ser. EPA-600/3-88-003, 91 pp.

–., and M. A. Perkins. In press. Oxygen deficit rate as a trophic state index. *J. Water Poll. Cont. Fed.*

–., and D. E. Spyridakis. 1974. Phosphorus cycle in Lake Sammamish. Coniferous Forest Biome Report, 4 pp.

–., and T. A. Wojtalik. 1968. Some aspects of increased water temperature on aquatic life. Tennessee Valley Auth., Chattanooga, Tenn., 48 pp.

–., J. A. Buckley, and R. M. Bush. 1972. Dilution as an algal bloom control. *J. Water Poll. Cont. Fed. 44:2245–65.*

–., G. R. Hendrey, and R. K. Stoll. 1975. Nutrient supply and the production and biomass of algae in four Washington lakes. *Oikos 26:47–54.*

–., C. A. Rock, and J. D. Krull. 1973a. Long-term lake recovery related to available phosphorus. In *Modeling the Eutrophication Process*, E. J. Middlebrooks et al. (eds.), pp. 5–14. Utah State University, Logan.

–., R. M. Bush, D. E. Spyridakis, and M. B. Saikewicz. 1973b. Alternatives for eutrophication control in Moses Lake, Washington. Dept. of Civil Engineering, University of Washington, Seattle, 102 pp.

–., M. A. Perkins, D. Lynch, and P. Hufschmidt. 1979. Internal phosphorus related to rooted macrophytes in a shallow lake. Proc. of a Workshop/Conference on Efficacy and Impact of Intensive Plant Harvesting in Lake Management. Madison, Wis., February 14–16.

Wetzel, R. G. 1964. A comparative study of the primary productivity of higher aquatic plants, periphyton, and phytoplankton in a large shallow lake. *Int. Rev. Geo. Hydrobiol. 49:1–64.*

–. 1972. The role of carbon in hard-water marl lakes. In *Nutrients and Eutrophication: the Limiting Nutrient Controversy*, G. E. Likens (ed.) Spec. Symp., Amer. Soc. Limnol. Oceanog. *1:84–91.*

–. 1975. *Limnology.* W. B. Saunders, Philadelphia.

–., and R. A. Hough. 1973. Productivity and role of aquatic macrophytes in lakes: an assessment. *Pol. Arch. Hydrobiol. 20:9–19.*

Wezernak, C. T., D. R. Lyzenga, and F. C. Polcyn. 1974. *Cladophora* distribution in Lake Ontario. Ecol. Res. Ser. EPA-660/3-74-022, 84 pp.

Wiederholm, T. 1974. Bottom fauna and eutrophication in the large lakes of Sweden. Ph.D. dissertation abstract, University of Uppsala, Sweden, 11 pp.

Wile, I. 1975. Lake restoration through mechanical harvesting of aquatic vegetation. *Verh. Internat. Verein. Limnol. 19:660–71.*

Wilhm, J. 1972. Graphical and mathematical analyses of biotic communities in polluted streams. *Ann. Rev. Entomol. 17:223–52.*

–., and T. Dorris. 1968. Biological parameters for water quality criteria. *Bio. Sci. 18:477–81.*

Williams, L. G., and D. I. Mount. 1965. Influence of zinc on periphyton communities. *Am. J. Bot. 52:26–34.*

Winter, D. F., K. Banse, and G. C. Anderson. 1975. The dynamics of phytoplankton blooms in Puget Sound, a fjord in Northwestern United States. *Marine Biol. 29:139–76.*

Wissmar, R. C., J. E. Richey, and D. E. Spyridakis. 1977. The importance of allochthonous particulate carbon pathways in a subalpine lake. *J. Fish Res. Bd.,* Canada, *43:1410–18.*

Witting, R. 1909. Zur Kenntnis des vom Winde erzeugten Oberflächenstromes. *Ann. Hydrogr.,* Berlin, *37:193–203.*

Woelke, C. E. 1968. Application of shellfish bioassay results to the Puget Sound Pulp Mill pollution problem. *Northwest Sci. 42:125–33.*

Wojtalik, T. A., T. F. Hall, and L. O. Hill. 1971. Monitoring ecological conditions associated with wide-scale applications of DMA 2, 4-D to aquatic environments. *Pesticides Monitoring J. 4:*184–203.

Woodwell, G. M. 1970. The energy cycle of the biosphere. *Sci. Am. 223 (3):*64–97.

Wright, J. C. 1958. The limnology of Canyon Ferry Reservoir. *Limnol. and Oceanog.* 3:150–9.

Wright, R. F., T. Dale, E. J. Gjessing, G. R. Hendrey, A. Henriksen, M. Johannessen, and I. P. Muniz. 1976. Impact of acid precipitation on freshwater ecosystems in Norway. *Water, Air and Soil Poll.* 6:483–99.

Zand, S. M. 1976. Indexes associated with information theory in water quality. *J. Water Poll. Cont. Fed. 48:*2026–31.

INDEX